LOOKING INSIDE THE BRAIN

LOOKING INSIDE THE BRAIN

THE POWER OF NEUROIMAGING

DENIS LE BIHAN

TRANSLATED BY TERESA LAVENDER FAGAN

PRINCETON UNIVERSITY PRESS
PRINCETON AND OXFORD

Library of Congress Cataloging-in-Publication Data
Le Bihan, D. (Denis), 1957–, author.
[Le Cerveau de cristal. English]
Looking inside the brain : the power of neuroimaging / Denis Le Bihan ; translated by
Teresa Lavender Fagan.
p. ; cm.
Originally published in France under the title Le Cerveau de cristal. Ce que nous révèle la
neuro-imagerie, copyright © Odile Jacob, 2012.
Includes bibliographical references and index.
ISBN 978-0-691-16061-0 (hardcover : alk. paper)
I. Fagan, Teresa Lavender, translator. II. Title.
[DNLM: 1. Neuroimaging—methods. 2. Brain—physiology. WL 141.5.N47]
RC78.7.N83
616.07'548—dc23   2014036587

British Library Cataloging-in-Publication Data is available

This book has been composed in Garamond Premier Pro and Trade GothicBold Condensed

Printed on acid-free paper. ∞

Printed in the United States of America

1 3 5 7 9 10 8 6 4 2

CONTENTS

CONTENTS

ACKNOWLEDGMENTS

This book would never have existed without the faith and enthusiasm of Odile Jacob, who literally forced me to put my ideas on paper, whereas I was extremely reluctant to do so, mainly because I was so pressed for time. Her team has done a remarkable job. I would like to thank in particular Nicolas Witkowski, who helped me to improve the text, both in terms of style and concision (it seems I have a slight tendency to digress), and who was able to give me the motivation and a rhythm that enabled me to complete the book. It has also been a pleasure to work with Teresa Fagan for the English translation and Eric Schwartz at Princeton University Press for the U.S. edition.

My life's work, upon which this book is based, has been undertaken over a period of around thirty years. Although a great part of my research has been undertaken in France, at the Hôpitaux de l'Assistance Publique in Paris, then at the Commissariat à l'Energie Atomique (CEA), first in the Service Hospitalier Frédéric-Joliot (SHFJ) in Orsay, and currently at NeuroSpin in Saclay, I have particularly warm memories of the years I spent at the National Institutes of Health in Bethesda, Maryland, and at the University of Kyoto.

As for my "prehistory," I owe a great deal to the support and understanding of Marie-Germaine Bousser, Emmanuel Cabanis, Maurice Guéron, Claude Marsault, as well as Denis Lallemand and Maurice Laval-Jeantet, who are no longer with us, who enabled me to simultaneously pursue a medical internship and studies in physics, a period during which I established the foundations of diffusion MRI. I wish also to thank the teams of Thomson-CGR for welcoming me into their fold, in particular Éric Breton and Patrick Leroux.

Throughout the years I have lost count of the many colleagues with whom I have had the joy and the honor to work. Unfortunately, I can't list them all, but I am particularly grateful to my National Institutes of Health (NIH) colleagues, John Doppman and Edwin Becker, who invited me, Robert Turner, Peter Basser, Jeff Alger, Robert Balaban, and Leslie Ungerleider, for all that they taught me. It is clear that the period I spent at the NIH corresponds to a sort of universal golden age in neuroimaging through MRI, with the appearance of fMRI, the groundbreaking developments of diffusion MRI (which

I brought back to France with me), which gave birth to diffusion tensor imaging at the NIH. At the University of Kyoto, diffusion MRI experienced new growth thanks to the unique climate in which I was working. I had, for example, particularly fruitful both scientific and friendly discussions with Hidenao Fukuyama, Toshihiko Aso, and Shin-ichi Urayama at the Human Brain Research Center, and with Kaori Togashi and Mami Iima in the radiology department.

As for André Syrota, his loyalty and constant support throughout these thirty years are indeed priceless. One can only admire what he has made possible over those years. After welcoming me at the SHFJ while I was working on my thesis, he invited me to return ten years later, after my stay at the NIH. And it was he who gave his immediate support to the project I proposed to him for a center of neuroimaging through high-field MRI, which was to become NeuroSpin. He was able to remove countless obstacles and convince the decision-makers, both at the CEA and in the highest levels of the French government. Bernard Bigot, the high commissioner, then chief administrator of the CEA, has also played and continues to play a crucial role in the creation and the development of NeuroSpin and its instruments—in particular, its future 11.7 T magnet.

Much of the work discussed in this book is of course the fruit of the labor of my close collaborators, Cyril Poupon, Stanislas Dehaene and his wife Ghislaine, Andreas Kleinschmidt, Christophe Pallier, Jean-François Mangin, Yann Cointepas, Bertrand Thirion, Lucie Hertz-Pannier, Jessica Dubois, Philippe Pinel, Jean-Baptiste Poline, Franck Lethimonnier, Luisa Ciobanu, Alexis Amadon, Nicolas Boulant, Sébastien Mériaux, Fawzi Boumezbeur, Éric Giacomini (and the list is, in fact, much longer). Their contributions have been enormous, and I thank them wholeheartedly, as I do my students, both past and current, of whom there are too many to mention here. The Iseult project has also been the occasion to create many strong bonds, both professional and of friendship, with our colleagues at the Institut de recherches sur les lois fondamentales de l'Univers (IRFU) at the CEA, notably the teams of Philippe Rebourgeard and Pierre Védrine, who have been involved in the conception and development of our 11.7 T magnet at the IRFU, and with the teams of Guerbet, led by Claire Corot and Philippe Robert for molecular imaging.

Finally, I can't express warmly enough all of my gratitude to my family, who have always believed in and supported me. First, my parents, who always encouraged me, and to whom I regret not having been able to give more back in return through my work; my daughters, Armelle and Carolyn; and my wife, Christiane, without whose constant love, understanding, and encouragement none of this would have been possible. This book is also theirs.

LOOKING INSIDE THE BRAIN

INTRODUCTION

It may be surprising that it has taken *Homo sapiens* centuries of reflection before realizing that the brain is the source of thought and awareness, centuries to allow the brain to usurp the organs—the heart or the liver (and its bile)—that had always been considered the centers of emotions. The Egyptians took the brain out of their mummies because they considered it to be merely a radiator. Aristotle, too, thought that the brain served to cool blood that was overheated through an emotional agitation of the heart, a concept that we find in the French expression *perdre son sang-froid*—"to lose one's cool." And yet the brain being so highly protected by the bone of the skull should have put people on the right track much earlier: it is clearly not by chance that nature provided the seat of our nervous system with such armor. Of course, traces of trephination discovered in prehistoric remains suggest that some curious observers had perhaps begun to suspect the truth. But for centuries the skull was associated more with death and its representations than seen as a container sheltering the organ of our thoughts.

And so it wasn't until the relatively recent age of the advent of anatomy, then surgery, that people dared to examine the brain, and to begin to study it. Much of our current knowledge comes from the work of Paul Broca on language in the mid-nineteenth century; from dissections of the brain of deceased patients; and from observing the brain of patients awake during neurosurgical interventions. Understandably, the possibility of seeing the brain of a living patient, let alone the brain of completely healthy subjects, for a long time remained an inaccessible dream. Only science fiction novels and films dared to suggest that one day it would be possible to have direct access to the contents of our thoughts by observing the human brain.

But this didn't take into account the progress that had been made in the realm of physics. Atoms had been discovered, and mysterious rays, such as

1

x-rays or those emanating from natural radioactivity, had been demonstrated. Following these discoveries theoretical and experimental physicists at the beginning of the twentieth century in a few years put into place physical and mathematical models to explain such strange phenomena and to account for the infinitely small. And so quantum physics was born, a discipline that explained the behavior of atoms and particles, the ultimate components of matter. Armed with these powerful theoretical tools, physicists were able to go beyond what was available in nature: radioactivity became artificial, atomic systems were mastered to produce a new, very powerful light—the laser—and, based on an equivalent principle, radio waves following the magnetization of certain atomic components. Engineers consolidated all these discoveries into developing apparatuses, such as x-ray tubes, enabling them to be reproduced on demand in well-controlled conditions, and to be used for various applications.

This is how physics joined the field of medicine, and in particular the field of medical imaging. The rays and particles discovered by physicists in fact had the ability to travel through our bodies, providing shadows of the structures they had traversed. Radiology made its debut as a medical specialty, and then came "nuclear" medicine. On a medical level, such images of the interior of the human body constituted an initial revolution, but those images remained fairly rudimentary, the organs were more "guessed at" than truly observed. The second revolution came with the marriage of physics and computer science at the end of the last century. Sensors of rays and particles became highly sensitive, and the computer was able to enhance that sensitivity, showing for the first time details of internal organs, and in particular of the brain inside the skull. "Neuroimaging" made great strides in the last quarter of the twentieth century, when it became possible to observe easily and without risk the brains of sick or healthy subjects, including that of a fetus in its mother's womb. Neuroimaging then became functional, revealing the workings of our brain while we are thinking, sometimes enabling access to the contents of our thoughts, or to the thought processes of which we are unaware.

How did neuroimaging become an incomparable method both for the neurosciences and for medical practice? What of our brain, and its functioning, do we really see in these images? How can we detect the signs of illnesses, both neurological and psychiatric, read our thoughts, gain access to our unconscious? These are the questions we will attempt to answer here. After recalling what our brain is, exactly, and why the concept of imaging is particularly adapted to it, our travels will lead us progressively from the instrument of imaging to what it enables us to see. For neuroimaging is essentially multidisciplinary: physics, instrumentation and engineering, software and sophisticated calculation processes enable that which otherwise would remain invisible to appear before our eyes; the biological sciences contribute, as well, whether it be molecular or integrative biology, neuropsychology or the cognitive sciences, neurology, psychiatry . . . The contributions of these disciplines are most often presented in their historical context, in order to show how science is made and unmade, depending upon human passions or dramas, Nobel Prizes or frustrations. Some chapters are more technical (chapters 1 and 3), but the chapters that follow them provide a glimpse into the huge realm of applications. Neuroimaging is well on its way; it has already changed our society and raised ethical questions. How far can it go? And will it enable human beings, themselves, to understand their own brains?

ONE

ELEMENTARY PARTICLES

Before setting out on our discovery of the brain, we must first understand what makes it such a unique organ (figure 1.1). The brain weighs on average close to 1.5 kilograms (3 pounds, for males, a bit less for females) and is made up of two hemispheres, the left and the right, which play somewhat different roles. It is part of a larger whole, the encephalon, which also includes the brainstem, through which neurons travel to communicate with the spinal cord (some neurons that start at the brain and end at the base of the spinal cord are more than a meter long . . .). Although relatively small in size, the brainstem is crucial because it gathers together vital smaller elements that regulate our life (sleeping, breathing, heartbeats). This is no doubt why it is very protected and difficult to access, because if it is damaged, the result is often fatal. Alongside the brainstem is the cerebellum (the "little brain," which in fact has little in common with the brain). One of its noticeable roles is to smooth out and coordinate our movements; thanks to the cerebellum we are able to walk straight or play the piano.

The human brain is impressive, perhaps less in its size (the brain of an elephant, whose memory is legendary, is larger, at 4 to 5 kilograms [11 pounds]), than in its complexity: the human brain is made up of a large number of folds and bumps, called sulci and gyri, that are not as developed in other animal species, including the great apes. It is to the French surgeon, Paul Broca, that we owe the fundamental discovery that the two hemispheres are not functionally identical, and that the brain is an organ that, although apparently homogeneous, is made up of regions that have different functional specificities, which is not the case with other such organs (the cells of the liver all do the same work regardless of their location in the organ). This discovery opened the door to modern cerebral physiology and is at the very heart of the goal of neuroimaging: more than just producing images, neuroimaging involves the creation of maps that show the "natural geography" of these regions, and, more important, how they are implicated in the sensory-motor or cognitive functions that they underlie.

BROCA'S DISCOVERIES

Paul Broca was a surgeon at the Bicêtre hospital (reference 1.1). In 1861 he had a patient, Monsieur Leborgne, who became unintentionally famous. In the hospital M. Leborgne was nicknamed "tan-tan" because he responded to questions (what's your name? what day is it? and so on) with only one syllable, "tan," which in general he repeated twice. This patient had what today is known as aphasia, a sort of "mutism," a disorder affecting the production of language, although he suffered from no paralysis of the "bucco-phonetic" muscles used in speech. After the patient died on April 17 of the same year, Broca did an autopsy and dissection of his brain. He found a lesion on it and drew two major hypotheses from this (figure 1.2), which the next day he presented to the Paris Anthropological Society: the patient's functional disorder must have been owing to the specific localization of the lesion in the brain. That lesion was toward the front, at the base of the third circumvolution of the frontal lobe, in the left hemisphere. Had the lesion been elsewhere, farther back, or in the right hemisphere, M. Leborgne would have perhaps had other symptoms, but he would not have been aphasic. This hypothesis was quickly tested on other patients. Thus was born the principle, which today has been well verified, of a direct link between cerebral localization and function, each cerebral region being associated with a specific function (motor functions, vision, hearing, language, and so on), and this all fitting together on different scales like a set of nesting Russian dolls.

The region that was affected in M. Leborgne's case is today known as Broca's area. Though the role of Broca's area in the production of language is beyond any doubt, we now know that many other cerebral areas (forming a network) are important to language. And inversely, Broca's area is also implicated in other functions. But it remains true that the postulate of regions of the brain being associated with localized functioning is completely established today.

Broca's second great discovery was that the two hemispheres each have their own areas of specialization: language is mainly seated in the left hemisphere (the twenty or so aphasic patients of Broca all had lesions on the left). Until Broca, the two hemispheres were assumed to have identical functions, like our two kidneys or two lungs, which have exactly the same role. For the first time, then, it appeared that the two cerebral hemispheres were not identical, not functionally interchangeable. In around 85 percent of us, the functions tied to language are predominately located in the left hemisphere; for the others, they are in the right hemisphere, and sometimes the dominance is less clear, with the two hemispheres clearly participating to the same degree in the production of language. At what moment, during the millions of years of evolution, did the lateralization of the human brain appear? Is this predisposition of the left hemisphere for language of genetic origin? Is it present in the brain of the fetus and the baby before they begin to speak? Is it linked to manual dexterity? Studies in neuroimaging are beginning to provide rudimentary answers to those questions.

Following Broca's discovery there was considerable progress, because this concept had opened a true breach in our understanding of the functioning of the brain. For more than a century neurologists (and in particular those of the French school at the beginning of the twentieth century, with Pierre Marie, Jules Déjerine, Joseph Babinksi, and many others) learned a lot from their patients with cerebral lesions. They had only to closely observe the patients and their functional disorders, then "recover" their brains after their deaths to establish a link between the localization of the lesion and the functional deficit. We owe a debt of gratitude to these neurologists, excellent brain sleuths, who, from their careful and detailed observation of sometimes tiny neurological signs (a small anomaly in an eye movement, or subtle cognitive disorders brought to light through complex tests), were able to establish the localization of the lesion within a few centimeters (even a few millimeters in the brainstem). Even if the nature of the problem (for example, the blockage of a small cerebral artery by a blood clot) could sometimes be suggested by symptoms, this rarely resulted in the patient being cured, as therapeutic treatments at that time were quite limited. This conceptual approach in which functional deficits were associated with cerebral localizations none-theless had its limits. First, brains had to be retrieved and dissected—and not all patients died! And those that did were not systematically autopsied. In addition, the localization of the lesions was not scientifically controlled in advance but were the work of Mother Nature. Whereas some regions were very often afflicted, others almost never were, and their functional role remained unexplored. This is explained in part by the fact that many lesions are of vascular origin, and that some vessels are more exposed or more fragile than others.

THE BIRTH OF MODERN NEUROIMAGING

A radically different approach was taken as neurosurgery began to advance in the 1950s, in particular within the school of the Canadian neurosurgeon Wilder Penfield. During surgical interventions on the brain patients were awakened during the operation (the brain, even though it is the primary nerve center, is not sensitive to pain). By touching or electrically stimulating a cerebral region, the patient could report directly on his or her sensations, such as: "my thumb is asleep." Personalized functional maps of the brain could thus be drawn for each patient to pinpoint the regions to be avoided during the removal of a cancerous tumor, or the source of epileptic seizures, in order to preserve the patient's motor or language functions. This approach for the first time enabled Broca's theories to be proven "positively" (the *expression* of functions), whereas up until then the process was "negative" (a *deficit* of functions in the patients with cerebral lesions), by eliciting expressions of the functional content of healthy cerebral regions.

It was within this context that in the 1970s modern, computerized neuro-imaging emerged, a true revolution that forever changed our approach to the brain. Until then neurologists had only cranial radiography at their disposal. X-rays, discovered by Wilhelm Röntgen in 1895, at best enabled the creation of shadowgraphs of the skull upon radiographic film, the shadows being roughly classed into four types depending on their intensity: bone and calci-fied structures (very opaque in x-rays); water and tissues containing water; fat (not very dense in x-rays); and air (transparent). X-rays of the skull thus showed only fractures of the bone, the appearance of abnormal blood vessels, and sometimes invasive tumors on the skull, occasionally calcified tumors, but not much more. One could go further by injecting a liquid containing iodine (opaque to x-rays) in the veins or arteries of the patient to display the vascular tree. Beyond direct afflictions of those vessels ("aneurysm": swelling of a fragment of the artery; "angioma": an abnormal proliferation of vessels; shrinking from an atheromatous plaque; blockage from a clot; and so on), one could guess at the presence of otherwise invisible lesions, such as tumors, from the fact that they displaced the normal vascular architecture[1] (figure 1.3). One could also "make opaque" the ventricular cavities, open spaces in the center of the brain that contain cerebrospinal fluid (CSF), by injecting an iodized liquid or a bubble of air (gas encephalography) into the spinal cord at the lower back. By making the patient assume acrobatic positions (to the point of standing on his or her head), the bubble of air would travel to the cerebral ventricles, then into a specific corner of those ventricles. Besides the discomfort of the method and the horrible headaches that often followed, here too, only lesions that displaced or molded the ventricular cavities could be seen. Everything else remained discouragingly invisible.

On the functional side, neurologists were able to record the electrical activity emitted by the brain (electroencephalography, or EEG). Indeed, nerve impulses that enable neurons and brain cells to communicate among each other depend on the movements of ions (atoms that have lost their electrical neutrality) such as sodium, potassium, or calcium. The movements of these charged particles represent little electrical currents that create localized electrical and magnetic fields that can be detected and recorded at a distance using electrodes placed on the scalp. But the signals, whose localization remained very uncertain, above all enabled the detection of electrical occurrences that were abnormal in their intensity, as in the case of epileptic seizures, true cerebral "electrical storms," or in their absence in the case of tumors or a localized affliction of cerebral tissue.

THE FIRST REVOLUTION: THE X-RAY CT SCANNER

The lives and comfort of these patients (and their neurologists) improved dramatically with the appearance of the x-ray computed tomography (CT) scanner in 1972 (figure 1.4), thanks to the English engineer Godfrey Houns-

[1] Intracerebral angiography is widely used today, but with another objective, within an interventional framework and in connection with therapeutic practices: this is the injection in situ of materials to plug a hemorrhagic break, or of medicine to dissolve a clot, for instance.

field, who, along with Allan McLeold Cormak, received the Nobel Prize in Medicine or Physiology in 1979. This date marks a turning point: the introduction of computers into radiology (reference 1.2). The brain would finally become visible without having to open the skull. First, classic radiographic film was replaced by x-ray-capturing sensors linked to a computer. Thus, instead of the four nuanced shadows detectable by the eye of the radiologist, the computer, through the sensors, could see hundreds. And so nuances between shadows produced by a healthy brain and those coming from lesions, or even different structures of the brain, could then be detected. Above all, instead of projecting a single dimension of shadow, the brain could be scanned under dozens, even hundreds, of different angles. This was made possible by the extreme sensitivity of the sensors and the digitization created by the computer: the dose of x-rays needed for a projection was infinitely reduced compared to classic radiography, which allowed the number of scans to be increased while keeping radiation at a low level.

Cormak showed that it was possible to combine these multiple projections, adding them to the computer's memory, to reconstruct, point by point, the entire picture of what was shown by the x-rays of the skull and its precious contents. The last step was to transform the virtual image into a real image by projecting it onto a screen (this general principle, moreover, was later used in future imaging methods that did not use x-rays). In CT scanning, scanning by x-rays is carried out on a plane (by turning the x-ray scanner around the head of the patient); the image obtained is thus that of a "slice," perpendicular to the axis of the head. In that slice (figure 1.4), one sees cutaneous structures, the bone of the skull with the tiny details of its internal structure, such as the external and internal bony parts of the skull, but above all, and for the first time, the interior of the skull, that is, the brain, the intracerebral ventricles, of course, and any lesions that might be found there—without an autopsy, without dissection, without pain or injury. The patient just lies down for a few minutes in the scanner and his or her brain is virtually dissected into slices by an x-ray beam.

The revolution is complete, and not only on a technological level. Radiologists, used to simple shadows, see much more than they did with their own eyes: and so they must learn to regulate the levels of contrast in the image in order to enhance a given structure. One speaks of "windows" of contrast, for it is indeed a matter of seeing, with the human eye, only a small bit of the landscape seen by the computer, by choosing the position and the width of the windows. Depending on that choice, the appearance of images can vary enormously, some windows allowing the details of the bones of the skull to appear more clearly, others the cerebral structures. Most important, at the beginning practitioners had to learn to think in terms of slices and rethink the entire three-dimensional (3D) anatomy of the brain learned during long years in anatomy classes and during dissections. Atlases of "tomodensitometry" (another name for computerized x-ray scanning) were created. For neurologists it was also a great surprise, a dream come true, but also from time to time disillusioning when they sometimes discovered that the localization of the lesion revealed from an in-depth analysis of the symptoms of their patient was not at all the same as the one revealed by the scanner . . .

With x-ray CT scanning it became possible, in the treatment of a living patient presenting neurological symptoms, and thus when there was still time to prescribe a treatment, to reveal the presence of a lesion, localize it, and understand its impact on neighboring functional regions, and even sometimes to specify its nature depending on how it was shown in the image. One could also again inject an iodized liquid or another contrasting agent to accentuate the distinction between healthy and abnormal tissue. Such an agent, opaque to x-rays, is distributed in the vascular network, even into the smallest capillaries. A lesion with many vessels, such as a tumor, is thus clearly apparent (figure 1.4). One could also follow the evolution of a lesion in time, in particular to monitor its progression or shrinking following treatment. But the eye of the scanner sometimes sees too much, lesions that one doesn't know what to do with, that one can't explain, fortuitous discoveries during research into another lesion. And some visible lesions, alas, remain without precise diagnosis and remain "nonidentifiable objects," and above all without treatment.

In not much more than a dozen years the CT scanner began to appear in hospitals, and in another dozen it began to be systematically integrated into the healthcare system. Anecdotally, when I was a medical resident in 1980, it had been strongly recommended that we not speak of the CT scanner in our reports, so we would not be called doctors of science fiction! And yet, at about the same time that the CT scanner appeared, another revolution was brewing, that of magnetic resonance imaging, or MRI, the premises of which were published in 1973 in the journal *Nature* by the American chemist Paul Lauterbur (reference 1.3). It took this extraordinary technology even longer to penetrate the medical world, as some renowned scientists "didn't believe it." Lauterbur didn't receive the Nobel Prize in Medicine, by the way, until 2003, along with the English physicist Peter Mansfield.

Nuclear Magnetism

The physical principle of MRI is radically different from x-ray CT, because rather than x-rays, a magnetic field and radio waves are used. They still have something in common, however: the marriage of a technology that creates signals in order to expose a contrast between different biological tissues, and computer science, to reconstruct images from those signals, images that are again in the form of slices. An MRI scanner is also a cylinder in which the patient is placed lying down (figure 1.5). The heart of the system is a very large magnet that produces an intense magnetic field, several tens of thousands times greater than the Earth's magnetic field—the one that orients the needles of our compasses.

Why is such magnetic intensity necessary? Because we are dealing with nothing less than the magnetization of the nuclei of atoms. Whereas the x-rays of a CT scanner interact with the electrons that surround atoms, the MRI goes to the atomic nucleus. That nucleus is made up of particles—protons and neutrons. The simplest possible nucleus is made up of a single proton; it is the nucleus of the hydrogen atom, which will be the hero (along with the water molecule in which it is found) of the rest of this book (alongside the brain, of course). Placed in a magnetic field, the proton in question is magnetized. In physical terms, it is said to possess a "magnetic moment," that enables it to be oriented in the direction of the field, a bit like the way the needle of a compass reacts in the Earth's magnetic field. But the analogy stops there. We are in the world of the infinitely small where quantum physics rules,[2] a world very different from our own. In fact, protons can line up in the direction of the field, or in opposing directions, the number of protons in each orientation depending greatly on the intensity of the magnetic

---

[2] Quantum mechanics is the branch of physics that describes fundamental phenomena at work in physical systems on an atomic and subatomic scale.

Box 1. A Subtle Equilibrium

Nuclear magnetism derives from the movement of charged particles (positive or negative) within atomic nuclei—"quarks." A proton (nucleus of the hydrogen atom) contains three quarks, which give it curious properties. Unlike our compasses, the proton can in fact be oriented in two ways in a magnetic field—in the direction of the field, or in the opposite direction. At very low temperatures, all protons line up in the direction of the field in a stable position of "rest," but as soon as that temperature changes, protons, due to thermic agitation, gradually move into the opposite position, which for them also represents a certain stability. Indeed, the position of rest is very fragile, and it takes only a minor change for the magnetization of the proton to move into the other direction. The balance between the two positions depends on temperature and the intensity of the magnetic field. In the terrestrial field, at the ambient temperature or at 37°C, the temperature of the brain, one finds approximately the same number of protons oriented in each direction, the difference being that two protons in a billion are oriented more in one direction than in the other. The resulting magnetic moment (that is, the overall magnetization of all the nuclei) comes uniquely from that infinitesimal difference, because the protons in opposite orientations reciprocally cancel out their effect. With stronger magnetic fields, that difference is accentuated, and the resulting magnetic moment increases.

field, and the overall magnetization being determined by the difference in the number of protons in those two orientations (see box 1).

In a magnetic field of 1.5 tesla (the "tesla" is the unit of measurement of the magnetic field; 1 tesla representing 20,000 times the strength of the magnetic field in Paris, for example), this difference is 50 protons per 1 billion, and at 3 teslas, it rises to 100 per 1 billion, which is still weak, but begins to make a difference. The first MRI magnets operated at fields of 0.1 to 0.3 tesla, but most of the magnets intended for MRI scanners used in hospitals today produce fields of 1.5 or 3 teslas. Not all atomic nuclei can be magnetized. In particular, those made up of an even number of protons and neutrons do not have a magnetic moment, their magnetic effects are nullified within the atomic nucleus. Unfortunately, included in these are carbon ^{12}C and oxygen ^{16}O, nonetheless key atoms in living matter.[3] This is why we generally resort to the hydrogen atom for MRI, because hydrogen comprises two-thirds of the water molecule (H_2O), which is abundant in the human body: more than 80 percent of the brain mass is comprised of water. Since there are around 30 millions of billions of water molecules, it begins to matter even if only 100 per 1 billion contribute to the observable magnetic moment. MRI is thus above all a story of water.

THE NUCLEI ENTER INTO RESONANCE: FROM NMR TO MRI

What are we going to do with this magnetized water (via its protons), which nature graciously puts at our disposal? This is where the concept of nuclear magnetic resonance (NMR) comes in, a concept that was introduced independently by the Americans Felix Bloch and Edward Mills Purcell in 1946, which earned them the Nobel Prize in 1952. The idea that is hidden behind NMR is that one can voluntarily "force" the orientation of the magnetic moment of atomic nuclei by entering into resonance with them, and thus modify their overall magnetization, by giving them the necessary energy in the form of radio waves of a very precise frequency (see box 2).

NMR continues to be enormously useful in physics and chemistry because nuclei inform us about their local environment, on an atomic scale. Indeed, these nuclei belong to atoms found in molecules (water for the hydrogen nucleus, but also many other complex molecules that contain hydrogen, such as amino acids, proteins, or DNA, to cite only molecules of biological interest). This environment upsets the local magnetic field perceived by the nuclei (mainly due to electrons present in the molecules, also carriers of magnetic moments), which is translated by small variations in the frequency of resonance of the waves reemitted by the nuclei. By carefully analyzing these waves, we can identify the frequencies, and deduce the molecules in which the nuclei are found, in what quantity, indeed even in what part of those molecules, at what distance from other nuclei, and so on. Last, with NMR spectroscopy we can reconstruct the three-dimensional structure of molecules as well as their temporal dynamics, a technology that earned Richard Ernst the Nobel Prize in Chemistry in 1991. NMR spectroscopy quickly appeared as

---

[3] "Exotic" variations (isotopes) such as carbon 13 (^{13}C: 6 protons, but 7 neutrons) or oxygen 17 (^{17}O: 8 protons and 9 neutrons) are magnetizable, but they are found in nature only in trace amounts.

Box 2. Atomic Tuner

The equilibrium between two orientations can be upset by an effect of resonance with radio waves—this is the principle of NMR. Radio waves, like light, are electromagnetic, but they can also be described by an elemental particle, the photon, carrier of a very determined amount of energy depending on the frequency (like a "color" for visible light). When an oriented proton absorbs a photon that has the right quantity of energy, it becomes "excited," and shifts to the opposite orientation. For the proton, the frequency of resonance is 42.6 megahertz (MHz) in a magnetic field of 1 tesla. All of this of course involves a very large number of protons. In this process, the energy absorbed is in large part restored, again in the form of photons—that is, radio waves of the same frequency that one must capture by means of an antenna and a radio receiver. The signal received is then amplified. Initially, its intensity directly reflects the number of magnetizable nuclei present—in this case, the number of protons and thus water molecules. Because the wave frequency depends on the nuclei, one can simply set the radio receiver at the right frequency to change the nucleus—if it is no longer hydrogen (and water) that one wants to study— exactly like choosing one's favorite station on a radio! At 1 tesla, the frequency is 42.6 MHz for the hydrogen nucleus, 40 MHz for the fluorine nucleus, and 17.2 MHz for the phosphorus nucleus. After a short time (called "relaxation" time), nuclei are de-excited and return to their original orientation. The energy is dissipated in the medium and the signal disappears.

---

an extremely powerful technology for physicists, chemists, and biochemists, more recently for biologists, and ultimately for physicians.

The principle of MRI (imaging through NMR) took root much later in the mind of Paul Lauterbur, in a fast-food restaurant in Pittsburgh, and became concrete through his scribbling on a paper tablecloth. Lauterbur, who had started his career as a chemist at home with a "junior chemist" set, had become familiar with NMR during his military service, which, thanks to his chemistry degree, he fulfilled as a member of the scientific staff. In an attempt to be transferred to a laboratory other than the one where he had first been assigned (developing chemical weapons . . .) he claimed that he knew about NMR. Lucky for him—and for us—that he did. Since the frequency of the

---

Box 3. The Brain Practices Scales

Resonance is also at the heart of MRI imaging. Here, one chooses the frequency of a nucleus (in general the hydrogen nucleus), but the magnetic field is varied slightly in a given direction. This "gradient" in the field induces a gradient in the frequency of resonance along a given direction, a bit like how the notes of a piano keyboard go from low to high. In each spatial position there thus corresponds a given frequency, just as with each key of the piano there corresponds a note. If a chord of several notes is played, we can, by listening to the sound (and being a bit of a musician), make out the keys (notes) that were played in the chord, and even the respective intensity with which they were struck. This is exactly what is achieved with MRI: the radio waves reemitted by the brain are made up of a superposition of multiple resonance frequencies of protons, according to their localization in the brain. One can find these frequencies, and for each one of them, their intensity linked to the number of implicated protons, by decomposing the global signal received with a mathematical operation called "Fourier's transformation" (what our ear naturally does for piano chords!). To obtain an "image" (that is, the number of protons implicated for each position in a slice of the brain, and no longer a projection in one direction), it is necessary to carry out the operation many times, by varying the direction of the gradient in space within the slice. We can then combine these signals to obtain an image representing the magnetization of the protons at each point of the brain. This is the fundamental principle of MRI, although many variants and improvements are used today to advance the method.

---

resonance of waves emitted by the nuclei depends on the magnetic field, his idea was to have the magnetic field vary progressively and in a controlled space (we speak of "gradients" in the field) to determine the spatial origin of the radio waves emitted by the nuclei of the object under study: for each position along the gradient of the field there is a corresponding frequency (figure 1.5); we can then obtain a map of the magnetization of the nuclei at each point in space—that is, an image (see box 3).

The first objects Lauterbur studied were a shell that his daughter had found on Long Island, a pepper, and tubes filled with water. The article he then wrote was rejected by the famous journal *Nature* because the images were too blurry. Lauterbur fought with the editors, and the article, which became a great classic, was finally published. Lauterbur later wrote that the entire history of science of the last fifty years could be written with the articles rejected by the prestigious scientific journals *Nature* and *Science*. Moreover, the idea of these "gradients" in the field had already been introduced by the French scientist Robert Gabillard twenty years earlier, but his work had remained unknown. Lauterbur was unable to convince his university (New York University at Stony Brook) to patent his invention (which he called "zeugmatography") in order to market it, which was a huge strategic mistake (the MRI market was estimated at close to 5 billion dollars in 2011). It took him almost ten years to obtain funds from the American government to construct a first prototype. During that time other processes to encode the spatial origin of NMR signals in an object, and thus to create images, were developed, without having to "turn" the gradient in a great number of directions. In particular Peter Mansfield of the University of Nottingham borrowed Lauterbur's idea and improved it by introducing a process of very rapid encoding and localization of signals. It then took another fifteen years for that improvement to be routinely available, but his university had taken out patents, and Mansfield, who became rich, was knighted by Her Majesty.

We must emphasize the real weakness of the energy from radio waves that creates the resonance of atomic nuclei. Those waves have nothing to do with nuclear energy, even if they involve atomic nuclei, and also remain very inferior to those associated with x-rays and radioactivity. Since there is no radioactivity or radiation involved, and in order to avoid any confusion in the public's mind, our legislators proposed dropping the term "nuclear." Thus was born "magnetic resonance imaging," or MRI.

THE ANATOMY OF AN MRI SCANNER

The key to MRI is thus this "gradient," this spatial variation of the field, which enables us to record, to encode the spatial origin of NMR signals. It is obtained through coils of copper wire through which a very intense current passes in a very short amount of time. The position of these coils enables the superimposition on the main field (intense and very spatially homogeneous, created by the main magnet) of another, much smaller field, but one that can be varied in three spatial directions. A basic MRI scanner is thus made up of a large magnet (in general in the shape of a cylinder) within which are placed the coils, and antennae to emit and receive radio waves. It is an assemblage whose dimensions are optimized to diminish the related costs. In other words, in the machine there is only the room necessary to place the object of study—in this case, the head of the patient (figure 1.5).

In order to produce intense magnetic fields over a large volume, a specific technology is required. Permanent magnets (like the ones we find on our refrigerator doors) can produce only limited fields (at most one- or two-tenths of a tesla) and are rarely used for MRI. Therefore, one resorts to the technology of superconducting electromagnets. Superconducting materials, such as an alloy of niobium and titanium, enable the passing of very intense currents (several hundred amperes are necessary to reach fields on the order of tesla over large volumes) because they offer no resistance to the current if they are cooled to a temperature of $-269°C$, which can be done by plunging the magnet into a bath of liquid helium. Superconductivity is also a recent discovery in the quantum properties of matter (it has generated no fewer than twenty Nobel Prizes since its discovery by Kamerlingh Onnes in 1911). It enables the current to circulate without any loss of energy: provided the magnet is cooled in liquid helium, it can produce its magnetic field *ad eternam*. Thus, the field is present even when the scanner is off, which necessitates some precautions when someone enters the room where it is located.

Another fact about MRI scanners is that the room where the magnet is placed must be shielded with copper, a Faraday cage: as we know, a simple shield or screen (such as that in a concrete tunnel) is enough to ruin the reception of our car radios. It is not like a lead screen used to limit the *escape* of x-rays from an x-ray scanner beyond a room to avoid any unwanted irradiation; rather it is an electromagnetic screen that prevents radio waves from *entering* the room and polluting the MRI signals. Indeed, the energy in play is very weak, and the signals produced by the protons of the brain are of the same order, if not inferior, to those emitted by hertzian radio, television, or telephone waves, and so on: at 1.5 tesla, the waves emitted by hydrogen nuclei have a frequency of 64 megahertz (MHz), and at 3 teslas, 125 MHz—a range of waves similar to those emitted in radiophonic transmission. If we don't want to transform the MRI scanner into a very expensive radio, we must protect its antennae from neighboring waves in order to capture exclusively the waves emitted by the atomic nuclei of the brain—and not the greatest hits.

In practice, an MRI exam consists of sending small bursts of radio waves into the hydrogen nuclei of the brain, then "listening" to the signals reemitted by those nuclei (by storing them in a computer memory), while alternately pulses of current are periodically sent to the gradient coils to localize the signals within the brain, all of this being organized extremely precisely and rapidly (dozens, indeed hundreds, of times per second). It is thus a true musical score (one speaks moreover of a MRI *sequence*) that the computer must play with the electronic equipment of the MRI scanner, after programming by the technical team.

Those who have already undergone an MRI exam know how noisy the scan is. Indeed, the very intense magnetic field produced by the main magnet contrasts with the small fields of localization produced by the gradient coils and subjects them to very great force. They are then subjected to mechanical vibration, which generates a sometimes very intense noise against which the subject must be protected. The type of noise generated depends on the MRI sequence; it goes from jack-hammering to more musical sounds of variable frequencies, because the sequence is very rapid. Some of my colleagues have even managed, temporarily, to convert MRI scanners into true (but unpleasant) musical instruments.

The Crystal Skull

In short, MRI involves placing our precious brain in the center of this gigantic magnet so that the brain's atomic nuclei can be instantaneously magnetized. The technician in charge expertly plays with the magnetization of the hydrogen nuclei by means of appropriate bursts of radio waves, and captures the resulting waves emitted in return by those nuclei in the central memory of the scanner's computer. There, "reconstruction" software sorts the waves in function of their original localization (the computer knows the map of the magnetic field programmed by the technician). We then have a latent image, a map of the magnetization of the hydrogen nuclei of the brain, which remains only to be projected onto a screen.

The x-ray scanner had trained the radiologist to see many details in the structures of bones. Indeed, x-rays interact with the electrons of atoms, and bone is very rich in calcium which, with its twenty electrons, greatly inhibits x-rays. Soft tissue such as the brain, 80 percent of which is water, is almost transparent to x-rays (hydrogen has only one electron). In MRI, it is just the opposite: one sees "soft" tissue very well, since the signal comes almost exclusively from the magnetization of the hydrogen atom of water. Bone gives little or no signal because it contains very little hydrogen. With MRI, it is thus the skull that becomes transparent . . . a crystal skull that suddenly lays bare the brain it protects and whose secret it has guarded up to now, offering its deepest recesses to our indiscreet look (we will see later that this term is not inappropriate!) with unprecedented precision.

Fat, such as that contained in the scalp or between the internal and external walls of the skull, rich in hydrogen atoms, also emits a strong signal. The MRI image is indeed foremost an image of the density of the tissue's hydrogen atoms. The images are extraordinary, very detailed and very contrasted (figure 1.6). For the first time we can clearly distinguish the gray matter that covers the brain from the deeper white matter, a distinction that is very difficult to make with the x-ray scanner. The difference in the density of hydrogen (or in the amount of water) between these two tissues is, however, very small—less than 10 percent. What is the origin of this new contrast, then?

We must look to another of the MRI pioneers, Raymond Damadian. This professor at the State University of New York (SUNY) Downstate Medical Center recounts how he was suffering from stomach pain and that he feared it was cancer. He thus sought a means to reveal the cancer in his abdomen, without having to be opened up, by using NMR. At the beginning of the 1970s MRI had not yet been born, nor had the x-ray CT scanner. However, Damadian had an interesting hypothesis (reference 1.4). An NMR study of samples of cancerous tissue had revealed an intriguing phenomenon. NMR signals are ephemeral: after having been rattled by radio waves, the magnetization of hydrogen atoms of water in tissues "relaxes"—that is, it spontaneously returns to its initial value of equilibrium. The time of that return is called "time of relaxation T1."

Now, it seemed at that time that the magnetization of cancerous tissue returned more slowly to a state of equilibrium than did normal tissue, having a longer T1 than that of healthy tissue. Roughly, the time of relaxation T1 reflects the interactions between the magnetization of the nuclei under study and the chemical environment in which they are found. In pure water, the molecules don't have much to encounter and their T1 is very long (a few seconds). In biological tissue that contains a large number of diverse molecules, the T1 of water is around 3 to 6 times shorter, but cancerous tissue in general has a longer T1 (that tissue swelling with water due to edema), whence Damadian's idea to measure the T1 of the hydrogen nuclei of his stomach. In 1972 he applied for a patent for a machine based on that principle, but one that used a procedure of localization of the NMR signal through focalization and scanning of the area point by point, which was much different and much less effective than Lauterbur's method that led to MRI. Seven years later he obtained a first image with his scanner (which he called *Indomitable*— nothing to do with the French nuclear submarine of the same name), but this was after Lauterbur and Mansfield. However, Damadian always considered himself one of the inventors of MRI (his machine is on display at the National Inventors Hall of Fame in Ohio and his supporters were extremely disappointed to see that the Nobel Prize for the MRI was not granted to him in 2003). However, in the meantime he had founded the company FONAR to exploit his patent and make MRI scanners, fighting with other builders whom he accused of stealing his patent. He thus won $129 million in the famous lawsuit that he brought against General Electric in 1997.

Today we know that the lengthening of T1 in cancerous tissue is not specific to cancer, that lengthening translates rather as an inflammation of the tissue and its infiltration by water (edema), which infuses the tissue with more liquid, thus accounting for a longer T1. But it remains true that the incredible contrast that one can see in MRI images comes principally from the difference in the time of relaxation between the tissue, in function of its composition. Thus, the times of relaxation of cerebral gray matter and white matter are very different, due to their different compositions, even if we are today incapable of explaining exactly why that is so by means of physical or mathematical models. Images produced by MRI scanners do not depend so much on the tissue's content in protons (and thus in water), but rather on the time of relaxation of that tissue, T1, but also T2, another time of relaxation that translates the duration during which the waves emitted by the hydrogen nuclei remain synchronous, which also depends on the nature of the tissue.

The technician in charge of the exam can thus adjust and manipulate this contrast as needed by altering the magnetization and demagnetization of the nuclei.

The so-called white matter can easily appear grayer or whiter than the so-called gray matter! The names "white" and "gray" were first assigned due to the way the tissue looked under the microscope after it had been altered with the chemical fixers of the time—they lose all meaning with MRI. Other structures, such as the small blood vessels or the clusters of neurons at the center of the brain or in the spinal cord (called "gray" nuclei) (figure 1.6), appear clearly, with unequal precision, the resolution of images today reaching a fraction of a millimeter. Compared to the brain sections obtained through dissection, MRI images often appear even a bit more detailed, but with this very great difference: the owner of the brain has spent only a short time lying in the magnetic field of the scanner . . . and has left with his or her brain intact!

Thanks to the progress achieved in the technology of the antennae used in MRI, the gradient coils and the memory capacity of computers, images can be acquired in great numbers and very quickly, at the rhythm of several images per second, which enables us to study dynamic processes, such as the beating of the heart or the movement of the intestines. These images are now routinely obtained in hospitals (at present more than 30,000 MRI scanners have been installed around the world, around 30 for 1 million inhabitants in the United States), and several million exams are performed every year.

In seeing the fantastic images of the brain provided by MRI, it is easy to forget that they are not real sections, but indeed virtual sections emanating from an ephemeral magnetization of the water of the brain, revealed by its passage through an MRI scanner (upon exiting the scanner, the magnetization disappears immediately and entirely, of course), images of very subtle and sustained magnetization of the protons that make up its water molecules. But those protons and water molecules, by being magnetized infinitesimally and then relaxing, reveal many secrets of the structure and the functioning of our brains, as we will discover in the following chapters.

TWO

THE MAGNETIC BRAIN

Once out of the MRI scanner, the subject, either a healthy volunteer in-volved in research, or a patient undergoing a medical exam, has left be-hind a magnetic imprint of his or her brain. Rather, we are left, in fact, with a very large number of signals, waves reemitted by the protons of the brain, which have been stored in the memory of a computer. Those signals, sorts of latent images such as those on film in the era of silver photography which had to be revealed through some chemical processing, then must be transformed into images that can be viewed on a monitor. This phase of "processing"—we call it "reconstruction"—depends solely on software and is carried out very rapidly with the computers we have today, so that the images generally ap-pear in real time.

The Computer at Work

We are dealing, however, with a very sophisticated process, because at the same time the images are being produced, certain defects are eliminated. The number of images produced is often considerable: several hundred during a single exam. The radiologist is in fact inundated with images. The brain is cut up into a large number of thin sections (dozens or hundreds depending on their thickness) from different orientations, as MRI images can be obtained on different planes—transverse (as was the case for the x-ray scanner), or par-allel to the face, or even perpendicular, from the nose to the occiput (see figure 1.6). And each of them can be shown with varying degrees of contrast, such as T1, T2, or a combination of them, or even other contrast, "dedicated," for example, to the detection of blood vessels.

With MRI the "negatoscope" film viewer by which the radiologist was used to viewing x-rays thus becomes obsolete. Not only does the radiologist have to rapidly pull out and sort the pertinent images from among hundreds, but he or she must also manipulate their contrast in real time. One must be able to step back to understand the nature of a lesion, imagine its shape and its location in the brain. To do this, the computer plays an enormous role, enabling the synthesis and the sorting of thousands of bits of data produced by MRI in order to present "digested" views. Computer scientists and mathematicians, specialists in image processing, perform miracles here. Whereas much of the previous chapter was dedicated to physicists, this chapter belongs to these other specialists.

With the x-ray CT scanner the radiologist had learned to "read" cerebral anatomy in sections, mentally reconstituting the anatomical structures, either normal or pathological, in space. This was even more difficult in the time of radiography because it was necessary to reconstruct the anatomy using only the projection of shadows of the organs on radiographic film. But current medical imaging software gives the radiologist directly a three-dimensional view very close to the real anatomy. Only very close, because it is just a "magnetic brain"—only what gave a usable MRI signal is visible. The rest is quite simply nonexistent. If, for example, the patient has a dental implant, it can happen that, depending on the material from which it is made (if it contains iron), it becomes a little magnet in the MRI scanner (dental fillings aren't a problem, because they are mainly made up of mercury, silver, copper, and tin, all nonmagnetic metals). This impromptu magnet modifies the magnetic field, affecting nearby protons of water, such as those in the front of the brain. Their frequency will thus change so that they will no longer give any signal detectable by the radio receiver of the MRI scanner. They will be lost, body and soul, and absent from the process of reconstruction of the image. The result? Deformations, even a great void at the front of the brain, which must not too quickly be interpreted as the presence of a missing part of the brain that is being examined. They can also, if their frequency is not too altered, remain in the image but appear in the wrong place, for example behind the brain, or even outside it!

The anomaly is quickly identified if the presence of a dental implant is known in advance (this is one of the reasons why patients are carefully questioned before any exam). The air contained in the sinuses of the face or at the base of the skull can also disturb the magnetic field and create deformations in the images, because the "magnetic susceptibility" of air is very different from that of the water of biological tissues (figure 2.1). Some researchers have proposed filling the cavities with water prior to an MRI scan (it had been noticed that these deformations were largely reduced among patients with colds or sinusitis, as the air in those cavities is replaced by mucus). New MRI technology generally enables us to overcome these negative effects, but such anomalies in the magnetic brain sometimes remain subtle and difficult to distinguish from real pathology. A great classic case is the effect of mascara that patients may be wearing. The pigments used for the colorants sometimes contain iron, susceptible to magnetization and to destroying (virtually) a part of their . . . magnetic . . . brain. As for piercings, it's best to forego them.

The Brain's GPS

Once reconstructed, the three-dimensional images of the brain, which have an accuracy of around a millimeter, are used for a true "neuro-navigation." The computer first "peels" the skin from the face and the scalp (figure 2.2). These structures are very visible in MRI (but hair or a beard are not, because they do not "resonate," which means that reconstructed faces are not always easy to identify). The bone of the skull then appears, or rather the fat, rich in protons, which is inside it. Then, with a stroke of a keyboard or a mouse the bone disappears. We are now on the surface of the brain, which we can admire in all its folds, or rather its circumvolutions, and from any angle, above, below, sideways, and so on, to better highlight a given element that appears abnormal.

We can even ask the computer's virtual "scalpel" to make a cut in a suspicious place in order to see it in depth. The radiologist and even more important the neurosurgeon, thus, with the movement of a mouse, as in a computer game, have the innermost reaches of the patient's brain available to them. They can simulate how they will operate, see "where they must go" in a particular patient, so as not to harm blood vessels or other important areas that, when touched, could cause serious aftereffects. What is more, since many elements of neurosurgical procedures are today carried out by robots (the neurosurgeon of course manning the controls), it is possible to program very precisely the movements of the "hand" of the robot to prevent it from going into risky zones and to give it, within a millimeter, the coordinates of the zone to be reached, a bit like one programs the automatic pilots in airplanes. And so MRI has indeed become a true GPS for the neurosurgeon. But unlike a GPS that navigates a single Earth, here we must remember that all brains are different, which means the calculation of coordinates for each patient must be redone each time.

Of course, at first sight all brains look alike: two hemispheres, frontal lobes (in front), occipitals (behind), parietal (on the side, above and behind), and temporal (on the side, below) (see figure 1.1). The "bumps," gyri, and the "fissures," or sulci, such as the central sulcus or the so-called fissure of Rolando that separates the frontal, parietal, and temporal lobes, are present in everyone. But when we look carefully, we see that there are notable individual differences. The "calcarine fissure" in the middle of the visual areas in the occipital lobe has a complex shape, extremely variable from one person to another. These differences are clearly revealed with the help of computer technology. We can try, through specialized software, to superimpose two brains, by enlarging or shrinking them, lengthening or compressing them, turning or repositioning them, operations we call "replacement," but it remains impossible to make two different brains match up exactly. The position of the fissure of Rolando (in the motor-control region of the brain), for example, can vary by more than a centimeter from one person to another (figure 2.3). Another sulcus, in the parietal lobe, can appear to be divided into several pieces in some people, and continuous in others. These anatomical differences can be problematic for neurosurgeons, and thus the importance of an individual assessment of each patient cannot be overemphasized.

WHEN THE BRAIN IS CONSTRUCTED

To understand how the human brain can be both so similar yet so different among individuals, we must go back to the stages that presided over the creation and development of our brain during our mother's pregnancy. Since

MRI does not use ionizing radiation, like x-rays or radioactivity, it enables not only the examination of the same person many times to observe how his or her brain is evolving through time, but above all, it enables us to examine babies, fetuses, or even embryos in the mother's uterus to observe their brains, virtually—in particular, if an anomaly that might lead to a medical intervention is suspected. Ultrasonography is another very reliable imaging technology and is much less costly than MRI, but it does not enable us to see the interior of the brain with as much contrast and detail. In the beginning, at the embryonic stage, the future brain is a simple hollow tube aligned along the axis of the embryo. The hollow part then evolves to form the ventricular cavities. The wall of the tube manufactures a great number of neurons. During the final months of a pregnancy a simple calculation shows that the production of neurons reaches 250,000 per minute!

These exceedingly numerous neurons cannot remain where they were created, and so migrate to find a precise place on the surface of the future brain, "their" place (somewhat like vacationers who look for their favorite place on the beach), a place from which they will never move again. This overall migration process, minutely orchestrated by the genetic code, includes several stages—sequential in animals, and more intricate in humans. It is complex and still little understood.

The migrating neurons are assisted in their voyage by glial cells, which are four to six times more abundant than neurons. Their "logistical" support role seems more important than we have thought, in particular during the period of cerebral development. Indeed, the glial cells "show neurons the way" by weaving very fine "threads" along which the neurons migrate as they settle in the region of the cortex to which they are destined. What causes a neuron to go to one region or another (and thus to be assigned *in fine* to a particular function rather than another)? This remains a mystery.

This migration process sometimes meets with failure, certain neurons remaining blocked in their migration. Today, in particular thanks to the imaging that has for several years enabled us to identify them, we know that these very small clusters of isolated neurons, "ectopic" as doctors call them, sorts of islands in the middle of white matter, can be responsible for epileptic seizures in both children and adults, seizures that unfortunately occur quite frequently and can be debilitating, and are often resistant to treatment with medication. Depending on their localization, which is detected precisely through MRI, these clusters can possibly be removed through neurosurgery, leading to the elimination of seizures and thus a cure for the patient.

Today it seems that a number of afflictions that present in an adult actually originate during embryonic and fetal development. MRI is not, of course, the preferred imaging method to use in following a pregnancy, but it can be indicated in some difficult cases. This is why we have MRI images of the brains of fetuses or premature babies at different stages in their development (figure 2.4; reference 2.1). The magnetic brains reconstructed from MRI reveal the way in which the different structures appear, including brain lobes winding around the center, and how gyri are formed, in particular between the 26th and 36th weeks of the pregnancy. The center of the brain is sometimes—wrongly—called the "reptilian brain." In fact, our brain is not the result of an addition to more ancient animal brains, brain matter that was added in the course of evolution, but represents a continuum in which all the fundamental cerebral structures are more or less present from the start, though they develop differently depending on the species.

Mechanical factors, tension and stretching, must certainly interfere with the genetic code to give the brain its final form, but these factors and their interactions are still very badly understood. The approximate shape of the brain is moreover achieved only at the end of pregnancy, and premature babies are born with an immature brain. Girls seem to have a reduced cortical surface compared to boys, as well as a slightly inferior volume of gray and white matter, whereas the degree of folding is similar. Asymmetries between the left and right hemispheres become apparent early on.

It is also interesting to note that twins, who compete for space in the uterus, have a delayed, but "harmonious," cerebral development: the circumvolutions are less developed compared to a single fetus at the same age, but completely in proportion with a reduced cerebral surface. Everything eventually evens out, which is not the case for fetuses suffering from cerebral birth defects (reference 2.1). For them, the cerebral surface remains inferior to what the degree of folding would lead us to believe. But unlike other animal species, the brain of a small human is far from being completed at birth. It is not until the end of adolescence that the structure of the brain becomes essentially stabilized. Mother Nature (and her ally, Evolution) has once again done things well: humans don't have to endure lengthy pregnancies, and above all, most postnatal cerebral development occurs because of direct interaction with the environment, in conjunction with the neuronal and genetic legacy with which we are born.

THE DESTINY OF NEURONS

At birth the brain has approximately 100 billion neurons and weighs less than 400 grams, far from the three pounds it will weigh at an adult age, whereas many neurons will have disappeared by then. More than the quantity of neurons, it is the number of their connections that is important. Moreover, the second largest brain known to date was that of the notorious criminal Edward H. Rulloff, as was revealed at his autopsy in 1871. The brain of Einstein, however, was no larger than any other, but, on the contrary, smaller than normal (1,230 grams). It seems that several parts of Einstein's brain were missing or were hypotrophied, and others hypertrophied—in particular, the parietal lobes, which (among other functions) control mathematical reasoning and a sense of space. The study of Einstein's brain, moreover, cost Dr. Thomas Harvey, who was in charge of the brain's autopsy shortly after Einstein's death, his job and his reputation. He stole the brain, cut it into small cubes, then studied it at his leisure. Fired from Princeton University for his actions, he ultimately decided, many years later, to give the brain to Einstein's granddaughter, who was living in California. And so he traveled across the United States with Einstein's brain in a Tupperware container on the back seat of his Buick . . . and then brought it back to Princeton because Einstein's granddaughter wanted nothing to do with it. The brain, or rather what is left of it, is today preserved at Princeton University.

At birth, each neuron is connected on average to 500 of its kind. At the end of adolescence, each neuron can connect to 10,000 others. If we consider the total number of neurons, that means we have up to a quadrillion possible connections, and the number of combinations within a single brain is quite simply astronomical, which has led some to say that the brain is the most complex structure in the (known) universe.

And so it is not surprising that we are all different, including identical twins. It is to accommodate this considerable number of neurons and connections that our brain is so folded—the entire (unfolded) surface needs to be used! Neurons are in large part spread over an exterior area 2 to 4 millimeters thick, covering the surface of the brain, in six layers that form the cerebral cortex (figure 2.5). This is the "gray matter" that we have already mentioned, and neurons are sometimes called "gray cells," as opposed to white matter, which we will discuss later.[1] These neurons have variable shapes depending on where they are found in the cortex. Their morphologies are no doubt related to their localized function (hearing, vision, . . .), but the exact relationship that governs the emergence of an elementary function from this morphology and the connections to other regions, which could represent a sort of "neural code," remains unknown.

Neurons communicate among each other through filament extensions called "axons" (figure 2.5). The ends of axons are connected to other neurons through structures called "synapses," which contain chemical molecules, "neurotransmitters," such as dopamine, noradrenaline, or acetylcholine. These axons can be very short or reach several millimeters or centimeters in length, when they connect neurons located in noncontiguous regions of the brain—for example, those found in two different lobes—and can even reach up to a meter for neurons descending through the spinal cord. These long-distance connections, made of groups of axons, form the white matter that fills the brain wherever there is no gray matter. This "whiteness," visible on sections of dissected brains, comes from the fact that the axons are surrounded by a sheath of fat, myelin, which acts as insulation. But unlike our electrical wires, axons conduct the current better (faster) when their insulation (myelin) is thicker. There is also a very large amount of gray matter (and thus neurons) in the center of the brain, forming deep structures that have not migrated to the surface, basal ganglia, as well as in the brain stem, where it is responsible for vital functions and elementary behavior (feeding oneself, reproducing, sleeping, dreaming, or being awake and conscious . . .), and which thus in some way creates "the beast within us." These basal ganglia are quite visible (in "black") in an MRI image, especially as they often contain traces of iron (whose role is unknown), especially among elderly subjects (figure 2.6). These nuclei are affected, for example, in patients with Parkinson's disease: the neurons of these nuclei no longer produce the dopamine essential to the functioning of other neuronal circuits whose role is important for motor control. Anti-Parkinson's medication supplements the brain's failing production of dopamine.

---

[1] These terms "white" or "gray" historically come from the qualities observed by the first neurologists during their dissections. With MRI the contrast between "gray" and "white" matter no longer has relevance and can be completely inversed, as we saw in the previous chapter.

A fascinating procedure, reserved for certain young patients, consists of directly implanting an electrode in these nuclei to stimulate the neurons. The way this profound cerebral neurostimulation actually works is still not well understood. The positioning of the electrodes, determined from images obtained prior to the procedure, is critical, as the size of the zone (nucleus) to be stimulated is no greater than a few millimeters. And the procedure must be monitored with great precision, which means that the patient must be awake during the procedure so that the results of the stimulation where the electrode was temporarily placed (figure 2.6) can be observed. Thus some patients report having a very unpleasant sensation of acute depression with suicidal thoughts when an electrode has temporarily touched the region of the substantia nigra, just two millimeters below the subthalamic nucleus that was targeted. Such symptoms of depression following the implanting of electrodes might exist among around 12 percent of patients, but some depressed patients saw their symptoms improve after the placement of electrodes, which again emphasizes our lack of understanding of the intimate mechanisms of deep cerebral stimulation. In the future, the "sick" neurons could be stimulated not by an electrical field produced by electrodes, but by the light of a laser used in conjunction with implanted optical fibers. This promising technology is called "optogenetics." Following genetic manipulation, in fact, the membrane of the modified neurons, and those alone, can in fact become sensitive to light.

Language and Cerebral Plasticity

The noninvasive and harmless nature of MRI also enables us to examine a large number of healthy subjects to analyze variations in the structure and function of the brain over the course of a life. The images from these "cohorts" of dozens, hundreds, or thousands of subjects obtained in multiple sites, sometimes in several countries, can then be studied with powerful software in order to discover the traits they have in common. These images also enable us to observe variations that exist among individuals, but also the evolution of the brain and the effects of aging, both normal and pathological. Of note is the American Alzheimer's Disease Neuroimaging Initiative (ADNI) for the study of patients stricken with Alzheimer's disease, which now extends to Europe and Asia, and Imagen, a current European initiative aiming to understand the origins and effects of addiction on the brains of adolescents.

How does our brain evolve as we become older? If, at birth, we have all of our neurons, these gradually disappear as we age, following a completely normal process. It has been found that neurons do not regenerate, with the notable exception, perhaps, of those of the hippocampus at the base of the brain. The hippocampus is a complex structure that is involved, among other things, in short-term memory (figure 2.7). The neurons found in it are very sensitive to cortisol, a hormone secreted by the adrenal glands during stress, which literally "destroys" them. One might conclude that nature has once again done her job, as the destruction of these neurons through stress enable us to forget the conditions or serious events (accidents, war events, rape, and so on) that caused that stress. The destroyed neurons are then replaced by completely new neurons, ready for a new life . . .

Neurons die throughout the brain, and it is a completely normal process —we have too many of them! Furthermore, there are juvenile illnesses for which natural neuronal death is delayed. Although their gray matter is more abundant, these children show great cognitive and neuropsychological problems. In the first two years of life connections in the brain can develop at the considerable rate of 2 million per second, and the connections that have become useless are expelled. This natural, pitiless "cleaning" operation in which the neurons that aren't used (aren't connected) are eliminated by "sweeping" cells (microglia) must be done so that our brain functions normally. Neurons absolutely need a social life—to be connected—to live. But beyond this operation, we are the ones, through our genetic history and our environment, who ultimately decide the fate of our neurons: which ones we'll keep, which ones we'll eliminate, or, conversely, how we will connect them.

For example, it is well known that the Japanese have difficulty distinguishing the sounds "r" and "l": "long" and "wrong" sound identical to them, just as "election" and "erection," have no (phonetic) difference. And yet a Japanese baby knows the difference between "l" and "r"! How do we know? A baby's brain, whether Japanese, French, or American, is above all a "learning machine," very curious to learn all that is new. If we place an acoustic helmet on a Japanese baby's head and have him listen to "long, long, long . . ." and he has a pacifier in his mouth, nothing happens. If suddenly the sound becomes "wrong, wrong, wrong . . ." he begins sucking the pacifier more rapidly, a sign that he has perceived a change. After a year of life and exposure to a uniquely Japanese environment, and because the Japanese language doesn't include either "l" or "r," the ability to distinguish the sounds is erased from his brain since it is not being used (Japanese raised abroad or in bilingual families retain that ability throughout their lives). Americans exploited this particularity during World War II by cramming their secret codes with "r" and "l" in order to limit interception by the Japanese. The French have almost the same problem with the English sound "th," which they often pronounce "z"—the frequency bandwidth of the English sounds used for language is almost twice that of French.

But, conversely, the Japanese can distinguish sounds that are inaudible to us—in particular, long vowels (very frequent in Asian languages). If you ask a taxi driver in Kyoto to drive you to the temple of Honen-in, he won't know where to go. To visit that temple (magnificent, especially in autumn—it's worth the linguistic effort!) you need to tell him "Hoonen-in." The difference appears subtle . . . for us, but it is essential for the Japanese. To give an idea of the difference, let's listen to the English word "start," a word that poses no problem. But "start" is unpronounceable for the Japanese, who don't like adjacent consonants. Their brains will add imaginary vowels between the consonants, and "start" for them becomes (this is what they think they hear, and ultimately write) "su-taa-to" (note that the "r" has of course disappeared, replaced by a long "a" . . .).

What do our magnetic images reveal? Let's raise a corner of the veil (the way we can *see* what is happening in the regions of our brain will be the subject of the next two chapters). Certain cerebral regions—in particular, in the left temporal lobe near the auditory regions—used by a Japanese person to separate "Honen-in" from "Hoonen-in" (or "Kyooto" from "Kyoto," a very subtle difference for us) is the same that, for us, enables us to aurally distinguish "start" from "su-taa-to," whereas a Japanese person hears the same thing in these two words (his or her brain automatically "adding" a vowel after a consonant, and removing the unpronounceable "r") (reference 2.2). The presence of this region in our brain (and the still unknown mechanisms of the decoding of language that comes out of it), regardless of the language, surely has a genetic origin. But since the probability that we have genes coded for French, English, or Japanese pronunciation is almost nil, we must conclude that the functional structuring of this region, like that of our entire brain, is determined by its exposure to the environment. This cerebral plasticity, linked to the possibility of developing and stabilizing the connections between neurons, intervenes much more easily and rapidly when we are young—"you can't teach an old dog new tricks."

GENES OR ENVIRONMENT?

How does genetic programming enable our brain to be at the same time extremely sensitive to the environment and capable of evolving while modifying its own structure in the course of time? We have only around 20,000 genes (like the chimpanzee and the mouse; worms have merely half; and, let's be modest, whereas each cell of a simple grain of rice whose intelligence has not yet been revealed has more than 40,000 . . .). These genes can therefore not account individually for our 100 billion neurons and their 100 million billion connections. Our genes no doubt play a very large role in the macroscopic and microscopic structuring of our brain, and are probably responsible, for example, for the organization, the (still unknown) specific functional architecture of the neurons in the Broca area that cause it to be used by the brain for language. But no gene determines in advance whether that region is going to deal with French, English, Japanese, or Chinese. This linguistic specification is dictated by exposure to an environment, which induces profound modifications into the microanatomy of the brain, causing some parts to shrink and others to be enlarged. Genes are encoded for the biological structure that enables this plasticity, at least up to a certain point.

Can we imagine a computer that would be capable of modifying not only its own software (this already exists in part), but also its internal wiring, its circuits or its video card, depending on demand? This is one of the great mysteries we must solve as we attempt to understand our brain and potentially to repair it in the event of abnormalities, whether innate or acquired. To do that, we rely greatly on the mouse (not the one of the computer . . .), whose genome we know very well and whose brain we can see develop in the smallest detail through MRI. The magnets of the MRI scanners dedicated to this small animal can achieve very elevated magnetic fields due to the small size of mice, which raises the issue of some technical (and financial) constraints encountered in scanners intended for humans. With MRI scanners operating at a very high magnetic field (11.7, 14, or even 17 teslas), we can see

33

the interior of the brain of a mouse with microscopic precision, in particular during the phase of embryonic development. It then becomes possible to observe changes in the cerebral structure induced by controlled modifications of certain genes.

If, for humans, we are still far from knowing the specific role our genes play in the structuring of our brain, links have been established between genetic anomalies and cerebral dysfunction. Neuroimaging in fact shows that anomalies of genetic origin can be responsible for actual—but sometimes subtle—"malformations" in cerebral anatomy that can lead to functional disorders. This is the case with Turner syndrome. This genetic affliction, identified in 1938, is found only in girls (1 out of 2,000 births) and is characterized by the complete or partial absence of one of the two X sex chromosomes. The condition is characterized by various symptoms, among them below-average height and many anomalies during puberty. There is no mental retardation, and language in general is normal, but psychosocial problems occur, as do others involving mathematical calculations and visuospatial coordination. MRI images show anomalies in the upper temporal sulcus of the left hemisphere and the orbitofrontal cortex at the front of the brain, regions affecting social behavior, and anomalies in the right intraparietal sulcus, which is an important region for mathematical and spatial cognition (figure 2.8; reference 2.3).

However, anomalies in the anatomical structure of the brain can most often go unnoticed, even to the sharp eyes of an alert radiologist. Such anomalies create relatively subtle differences that only the sophisticated software of image processing can reveal, necessitating equally sophisticated means of calculation for the comparison of many subjects. When a blood sample reveals an abnormally high cholesterol level, this means that statistically the level is greater than in a population sampling. It can occur by chance, but there is a great likelihood that something isn't right in the subject's metabolism or nutrition. In imaging, up until now, radiologists didn't ask themselves this kind of question. They saw something clearly abnormal or they saw nothing at all. Thanks to computer science, this situation has evolved enormously. If the common strategy still consists in seeking effects over an average sampling, of obtaining a map of the brain of Mr. Everyone, we are becoming increasingly interested in deviations, in knowing whether the brain of Mr. or Mrs. Smith differs in any way from the standard brain. This is no minor affair, because we must first carefully define, identify, and agree on who will constitute the standard, so-called normal group. We can establish a "statistical atlas" by averaging the images of this group. Then, when a patient presents we can ask the computer to test, point by point (we talk of pixels, as with cameras or television screens) the presence of a statistical difference between the brain of the patient and that of the standard, "normal" brain, and ask it to show on a colored scale those pixels that appear abnormal. Thus regions that present too much or not enough gray or white matter can appear.

In the realm of pathology, in general we deal with a localized lack of gray matter, as the thickness of the cortical layer is reduced in a certain area. Thus, in some autistic individuals a diminution in the volume of gray matter has been found in the upper temporal sulcus (figure 1.1), which is involved in interpersonal relationships. Almost half the time anatomical anomalies, a priori of congenital origin, can be revealed, although we still can't firmly establish a direct connection between them. Among some schizophrenic subjects with auditory hallucinations, anomalies in the folding of the regions linked to language have been found. The effects of alcohol (which is toxic for neurons) can also be seen on MRI images: a study done on a sampling of chronic alcoholics in rehabilitation has revealed a significant decrease in the thickness of the cerebral cortex (gray matter)—in particular, in the frontal regions—and this especially when the consumption of alcohol began early in life (reference 2.4).

The Phrenology of the Brain

In the realm of physiology, we see rather a hypertrophy of certain regions. Everyone knows that London taxi drivers are among the best in the world (in any case, they're the ones who have been studied the most). They know London even into its remotest corners. This isn't so surprising when we consider that for many months they must travel the streets of London by bicycle before receiving their license. What is less known is that the "magnetic" brain of these drivers reveals that they have a larger hippocampus than that of the common person (figure 2.7; reference 2.5). We have seen that the hippocampus is the region at the base of the brain involved in short-term memory; it is through it that our memories penetrate into our brains. London taxi drivers have trained their memories to such a degree that their hippocampus has become hypertrophied. The difference is not enormous, of course, but sufficient to be detected by a computer. Bus drivers don't have a hypertrophied hippocampus, because they always follow the same route. This shows that it is not the fact of driving a lot that develops the hippocampus, but that of devoting oneself to tasks of navigation. By contrast, the hippocampus shrinks and atrophies among patients afflicted with Alzheimer's disease. It is, moreover, one of the first regions afflicted: one of the first signs of this illness is a loss of memory.

The same phenomenon of functional hypertrophy occurs in musicians: the motor cortex, the region that controls the hands, is generally more developed bilaterally in pianists than in nonpianists or even in other musicians (violinists or trumpet players) (figure 2.9; reference 2.6). Though a more subtle difference, musicians in general have more developed cerebral regions than does the common human. Thus the motor and sensory areas controlling the hand (especially left) are more developed in pianists, and the volume of matter (gray and white) in the Broca area is superior among orchestral musicians. In the auditory regions, such as the Heschl's gyrus, "synthetic" musicians (sensitive to the fundamental frequency of sounds) have more gray matter on the left, whereas "analytical" musicians (sensitive to the spectral composition of sounds) have more on the right. All these localized hypertrophies of the cortex reflect years of accumulated experience, but perhaps also predispositions ("gifts"). The planum temporale, for example, is *smaller* on the right among musicians with perfect pitch, suggesting a perhaps genetic origin (reference 2.7).

The concept that served as a basis for the phrenology of Franz Joseph Gall is thus quite real, and we indeed develop our brain in function of what we want to do with it, just as athletes develop some muscles more than others depending on their chosen sport. If phrenology went too far, it is because these hypertrophies are miniscule, just barely detectable by a computer, and insufficient to imprint their relief on the bone of the skull. The math bump does indeed exist, but we can't feel it while combing our hair.

What is even more surprising is the speed at which the brain reshapes itself. The younger we are, the more rapid the modifications, and it is clearly better to start learning piano at the age of 5 than at 70, even if one can learn it at any age. Regions of the brain linked to language, in particular, are still very plastic up to the age of 40 (reference 2.8). A recent study of volunteers learning to juggle shows that, in a few weeks, the brain is capable of changing its structure significantly—in particular, in the visual regions involved in the detection of the movement of objects, as is seen very well (by using a computer; the difference all the same remains very subtle to our eyes) by comparing MRI images acquired as the students were learning (figure 2.9; reference 2.9).

And so our brain at birth is a sort of clay to be modeled. It's up to us to make of it what we will. Up to now, images have enabled us to create the archaeology of the brain, to rediscover the traces of our personal interaction with the environment from the details of its anatomy, which more or less reveals the inner workings of our personal history. It becomes possible, with MRI images, to determine which regions, through their shape or, as we will see in the next two chapters, their function, have been determined by a given gene, and to see the effects of the environment—to determine the effects, in sum, of the innate and the acquired. These images reveal, for example, functional differences that are clearly visible over the entire brain among educated and illiterate individuals. Imaging reveals how education affects the functional organization of the brain, but also suggests ways to develop or optimize teaching methods adapted to the functioning of our brain, a fundamental issue for education. It is thus time to ask how it has become possible to see into the brain *live* and to plumb what is happening there in real time—for example, while you are reading this book. MRI from structural or anatomical has thus also become functional.

THREE

SEEING THE BRAIN THINK

Being able to access the contents of our thoughts has long been a dream of humankind: can we read our minds?

The answer isn't simple, but MRI enables us to see what is happening in a brain while a subject is thinking, and, under certain conditions, even to partially see into the contents of those thoughts. The ease with which images of brain activity can be obtained today (their interpretation is another story . . .) has turned them into a common tool for neuropsychologists. Before MRI, the relatively rudimentary images of brain function (and no longer simply the brain's anatomy) were obtained by nuclear imaging technology. That technology, like MRI today, rests on a fundamental postulate presented in 1890 by the English scientists Charles Smart Roy and Sir Charles Scott Sherrington (Nobel Prize in Medicine, 1932): active regions in the brain show an increase in blood flow in the vessels that supply them (reference 3.1).

I Think, Therefore I Irrigate

This phenomenon, called "neurovascular coupling," has largely been verified since then. In particular, the surgeon Wilder Penfield, who in his time (the 1950s) was considered to be the "greatest living Canadian" (and whom we mentioned in chapter 1—he awakened patients in the middle of surgical procedures), saw changes in color due to an influx of blood in specific regions of the brains of his (awake) patients on the operating table when they carried out certain tasks, such as moving their fingers. This influx of blood involved relatively large vessels on the surface of the brain, but we know today that the phenomenon also involves much smaller vessels, small feeding arteries, small drainage veins, and capillaries. And so the Swedish neuroscientist David Ingvar developed the idea of producing images of blood being supplied to the brain while his subjects carried out particular tasks, such as speaking or listening. The regions where the blood flow was high indicated regions that were affected by those tasks, such as the Broca area for producing language, and Wernicke's area for understanding language. This was 1974, and thus functional imaging was born, establishing the first direct link between psychology and neurophysiology.

As bizarre as it might seem, up to then psychologists had been interested in behavior, but not in the brain as an organ that they considered to be a black box. To obtain these images, Ingvar collaborated with the Danish physician Niels Lassen, who had perfected a completely new imaging method in the 1960s. This involved injecting a small amount of inert gas that had been rendered radioactive into the carotid artery. The passage of this tracer into the brain, showing cerebral circulation directly, was recorded by a scintillation camera placed at the side of the subject's head. The camera contained more than 200 detectors, each picking up the radioactivity emitted by around 1 square centimeter of the surface of the brain. Other results on cerebral blood flow were reported at the meeting in 1977, where some attendees voiced warnings against what appeared to be a new phrenology. Indeed, the method had several limitations of scale, such as the necessity of injecting into the carotid, and the rudimentary way images were obtained, since the detectors "saw" only a projection of the brain, without being able to separate the interior from the surface.

Seeing the Brain with Antimatter!

Considerable progress was made at the end of the 1970s with the introduction of the positron camera, or positron emission tomography (PET), or PET scan. PET appeared shortly after the invention of the x-ray scanner. This camera[1] is also a little marvel of physics (figure 3.1; reference 3.2). Radioactivity is still used, but what is injected, this time in a vein of the arm, is water

---

[1] The version of this camera for use with humans was invented by Edward Hoffman, Michael Ter-Pogossian, and Michael Phelps in 1973 at the University of Washington with the support of the Department of Energy.

that is made radioactive in a cyclotron (a particle accelerator). Whereas MRI relies on protons, the hydrogen nucleus of the water molecule, PET involves the oxygen nucleus of water. This nucleus is battered in the cyclotron until it loses a neutron, which turns oxygen O^{16} (natural oxygen, with eight protons and eight neutrons) into oxygen O^{15} (radioactive, with eight protons and seven neutrons). The nucleus then becomes very unstable and seeks an equilibrium, which it does by transforming one of its protons, which has become supernumerary, into a neutron, and by ejecting a positron—that is, an antielectron, a positively charged electron (this is indeed antimatter; we are at the heart of quantum physics). Oxygen (whose nucleus must have eight protons to be called oxygen) "loses its soul" here by now having a nucleus of nitrogen (whose nucleus has seven protons). The half-life of oxygen O^{15} is very short, around 2 minutes, which means that the nuclei prepared in the cyclotron will have lost half of their radioactivity at the end of 2 minutes, three-quarters at the end of 4 minutes, and so on. And so the cyclotron must be right next to the PET camera, and the radiochemists must work rapidly to incorporate this fleeting oxygen O^{15} into the water molecules, which must then be quickly injected into the waiting subject. Thanks to the circulation of blood, the injected radioactive water quickly reaches the brain, and in a quantity that is all the greater in the regions where we see an increased flow— that is, active regions.

It is in these regions that the positron intervenes. Indeed, the radioactive disintegration that leads to the emission of a positron is carried out in the brain. And this particle of antimatter has the same destiny as other particles of antimatter in the universe: it disappears in a ray of light as soon as it encounters a particle of equivalent matter—that is, an electron of cerebral tissue. It takes only a few millimeters of its journey for the positron to find its soul sister, and that constitutes the ultimate limit of the spatial resolution of the images obtained by this method.

With PET Einstein and his equation $E = mc^2$ are right at home: the light emitted corresponds exactly to the predicted energy; it is the result of the annihilation of the positron and the electron (which, by the way, means that a bit of cerebral matter, albeit infinitesimal, disappears in the process!). The atom from which an electron has been stolen must regroup by in turn stealing one from elsewhere . . . This is why we speak of ionizing radiation, but such radiation has minimal effects on health, as long as it remains within reasonable amounts. The light that is produced corresponds to the creation in two opposing directions of two photons (gamma rays) of very precise energy: 511,000 electron volts (eV). By way of comparison, it takes around 600 billion billion eV to light a 100-watt bulb for 1 second. One can easily imagine, then, that it would take some very special technology to "see" the photons issued from the disintegration of positrons in the brain.

THE GLORY OF PET

These photons are captured by extremely sensitive scintillator crystals spread around the subject's head, which "light up" as the photons pass, the effect being amplified with the help of photomultiplicators (figure 3.1). Using very sophisticated electronics, we target photons of 511,000 eV emitted at 180 degrees from one another, to ensure that they indeed come from the disintegration of a positron and not from the natural irradiation to which we are all constantly subjected, whether from the Earth itself, or from the depths of the universe. We can even calculate the temporal delay between the two photons to reconstitute (approximately) the place of emission within the brain ("time of flight" method). Radioactivity measurements are stored in a computer's memory and classified depending on their angle of measurement. These measurements are classified projection by projection, just as with an x-ray scanner. The difference, however, is that we are dealing with measurements of the *emission* of radioactivity (and not the *absorption* of x-rays by tissue), and that the source is found inside the brain (and not in an apparatus turning around the head). Last, after many corrections and other filtering, a map of the local radioactivity is reconstructed.

And so we are once again in the presence of sections of the brain, but they now represent the distribution of radioactivity (which reflects the level of cerebral blood flow) in each section. The anatomy is thus represented only indirectly—gray matter, for example, is distinguished from white matter because it is more vascularized. In addition, the precision of the images is only on the order of one centimeter, much less refined than those obtained through an x-ray scanner or MRI. PET has been the queen of functional imaging for close to 15 years because it has been the only imaging method that enables us to see activated cerebral regions during the execution of motor, sensory, or cognitive tasks. An entire scientific culture was born with the

41

appearance of PET, and the bases of modern functional neuroimaging have been established around these functional images: the manner of "stimulating" the brain of subjects, statistical methods for analyzing the images, anatomical readings of the activated regions, and so on. To be able to see in this way the brain while it is functioning had been a dream. For the first time we saw the regions of language "light up" when a subject spoke, or the visual regions when the subject looked at an image or a text (figure 3.2). This was also a revolution: we had gone from the static imaging of the anatomy of lesioned brains, from which a function had disappeared, to the search for a lesion with an x-ray scanner, to the imaging of the *functioning* of healthy brains among normal subjects using PET. Many studies were published on PET, focusing on where and under what conditions the brain was activated during the execution of a given task, and in an attempt to elucidate why and how the blood flow increased in the activated regions, following Roy and Sherrington's postulate of neurovascular coupling. The precise responses to these questions are still not known at present. In particular, PET studies have shown that the increase in blood flow is much greater than required by the brain's demand for energy (brought via the oxygen and glucose contained in the blood) when it is activated. Moreover, in a state of "rest" (in as much as that state has meaning), the brain alone consumes 25 percent of the organism's energy resources.

Positrons on the Verge of Being Replaced by Magnetism

The success of PET, as regards the imaging of cognitive functions, was ephemeral all the same because MRI showed that it, too, could show cerebral function. PET had some not negligible constraints, the first being that it relied on radioactivity, with isotopes with very short lives, necessitating the use of a cyclotron at the examination site. The radioactivity prevented the repeated use of a given subject, pregnant women, and children (except in cases of medical necessity, of course). Last, the precision of the images, in space and time, was of course exact, but far from optimal. The regions detected had to be a few centimeters in size and, taking into account the injected radioactivity, the images had to be produced at intervals of less than a minute. Nonetheless, PET is still used for cerebral functional imaging in specific cases, such as the study of cognitive functions linked to music. One advantage of PET is in fact that it is completely silent, whereas MRI is very noisy, not ideal for seeing how the brain reacts while listening to a Mozart sonata... In addition, the magnetic field of MRI does not allow musical instruments, such as an electric piano keyboard, to be introduced. Only PET enables us to see what is happening in the head of a pianist playing the Bach Italian concerto (the experiment has been done; reference 3.3). If PET is therefore rarely used anymore for cerebral functional imaging, it nonetheless remains a precious tool for other applications, and it is used more and more frequently in the study of neurotransmission and oncology, as we will see at the end of this book.

While PET was flourishing, a few MRI researchers were desperately trying to find ways for their favorite tool to do as much, and more. Everyone (including the author of this book) started from the principle of neurovascular coupling, which consists of creating images of cerebral irrigation in order to reveal activated regions in the brain. As other approaches were also envisioned (as we will see later) this demonstrated a certain conservatism: the scientific community is often hesitant, indeed actually reticent, to accept new concepts. At the beginning, then, scientists focused on obtaining images of cerebral irrigation through MRI. In 1986 I had obtained some preliminary images, but the results were very tenuous, uncertain, and not publishable. The first person to truly succeed, in 1991, was the American John Belliveau, of Massachusetts General Hospital in Boston (reference 3.4). He used a contrast agent that is used by radiologists to accentuate the native contrast (although it is naturally excellent) in images obtained through MRI. It must be said that certain lesions could remain undetected if their water content and the time of relaxation of the protons in these lesions is not very different from those of surrounding healthy cerebral tissue.

Electrons Come to the Aid of Protons

Here, too, conservatism prevailed: we had seen that radiologists increased the contrast of x-ray scanner images by injecting an iodized liquid into their patients, one that is more opaque to x-rays, in order to reveal the wealth of blood vessels in a tumor. Radiologists wanted to do the same in MRI—that is, inject something to reveal local vascularization. Again, Lauterbur came to their rescue with a chemical element: manganese (Mn).[2] Once administered into an organism, manganese profoundly modifies the contrast of tissues. This element in fact contains "single" electrons, so named because the electrons have a tendency to seek a soul mate and pair up. These single electrons, like protons, are bearers of a magnetic moment, but it is around 2000 times greater. In the magnetic field of the MRI, these electrons are also magnetized, and at the same time magnetize the atom of manganese that they comprise. We refer to "electronic paramagnetism" here, as opposed to the magnetization of atomic nuclei, "nuclear paramagnetism," that we have been discussing up to now.

Water molecules that are found near atoms of manganese thus experience a disturbed local magnetic field as compared to that produced by the magnet of the MRI scanner, which is translated by a slight modification of their own magnetization, and above all of their time of relaxation. In particular, this local disturbance of the magnetic field strongly accelerates the return to the magnetization of equilibrium after excitation through radio waves. Tissue containing manganese thus appears differently in images. In practice manganese, which is very toxic, is not used in humans. We use another atom, from the family of "rare earths," such as that of gadolinium (Gd), which also has single electrons (it is the chemical element that has the most).[3] Since gadolinium is also toxic, it is "hidden" within chemical complexes called "chelates," such as diethylenetriaminepentaacetic acid (DTPA) or DOTA (which corresponds to the nice name of acid 1.4.7.10-tetraazacyclododecane-1.4.7. 10-tetraacetic and not to a currently popular video game!). But water molecules succeed very well in reaching the gadolinium in the chelates and in sensing its magnetic influence, which accelerates their "relaxation." The complex "Gd-DTPA" is thus the contrasting agent that one injects into a patient's vein. Wherever it travels—that is, primarily in tissue rich in blood vessels—the magnetization of the molecules is modified, which can be seen in the im-

[2] Manganese and magnesium have similar names, as they both come from the Magnesia region in Greece, where they were first extracted and identified. This region in turn is named for the ancient tribe of Magnetes, who, in ancient Greece, were involved in the Trojan War.

[3] Gadolinium was identified at the end of the nineteenth century by the Swiss scientist Jean-Charles Gallissand de Marignac, then isolated by Paul-Émile Lecoq de Boisbaudran. But it was the Finnish chemist and geologist Johan Gadolin who (ultimately) gave the element his name, which is much simpler!

ages. A tumor, for example, appears much whiter than the neighboring tissue (figure 3.3). This procedure is performed routinely today in many patients for whom a brain MRI is prescribed to detect a lesion.

THE WRONG TRACK

Let's return to Belliveau and his attempt to use MRI to detect cerebral activation. The idea was simple: first, leave the subject at rest in the magnet, and then inject a small quantity of the contrast agent into him or her. The agent passes into the brain in a few seconds, and then is eliminated, while the images are being collected. By calculating the change in signal point by point, we can deduce how the magnetization has changed, and from there, with a few hypotheses, we can calculate the amount of contrasting agent that has passed through. Thus for each point in an image we can obtain a number that reflects the local volume of blood and its flow "at rest."

These calculations are done by a computer, and the final results are seen in the form of images in which contrast (or sometimes color) reflects cerebral irrigation. We then start over, but this time the subject looks (in the scanner, by means of glasses and prisms) at an image projected onto a screen—a sort of checkerboard whose squares alternate very quickly between white and black. This type of stimulus strongly activates the primary visual cortex. We thus obtain a series of images of cerebral irrigation in an "activated" condition. By comparing, point by point, the two sets of acquired images, we can obtain a map of the activated regions (those where the blood flow has changed), which appear in color on the background of the black-and-white "anatomical" images, to better localize the regions (figure 3.4). In this experiment, the back of the brain "lights up" in the area where the primary visual regions are found and where signals generated at eye level via the optic nerves arrive.

The publication of these results in 1991 in the American journal *Science*, which had devoted its cover to the article, exploded the world of neuroimaging. For the first time it was shown that MRI could also produce images of the activation of the brain in a normal subject, images that not only were much sharper than those obtained through PET but also that could be obtained without radioactivity or ionizing radiation. This began to disturb some members of the PET community. It must be said that the PET centers had invested quite a bit of money to acquire the costly cameras, not to mention the cyclotron, the radiochemistry lab, and the necessary staff. However, the worst for them was about to come . . . Belliveau's method was abandoned very quickly (less than a year later), because it was dethroned by a much more effective method.

A Question of Oxygen

A year earlier, my friend Seiji Ogawa, a Japanese researcher who at the time was working at the AT&T Bell labs in the United States (a multinational telecommunications corporation that was interested in biology!), had made a surprising observation: by comparing the images of the brains of rats obtained through MRI, he saw that they didn't look the same depending on whether the rats were breathing normal air or air deprived of oxygen. In particular, the blood vessels, which are scarcely visible in cerebral tissue under normal conditions, became very dark with the absence of oxygen (figure 3.5; reference 3.5). To explain the phenomenon, Ogawa referred to the work of Linus Pauling,[4] who had shown that hemoglobin possesses curious magnetic properties.

Everyone knows that the red color of blood comes from the presence of . . . red blood cells, which owe their color to a pigment, hemoglobin. The primary mission of this exceptional molecule is to transport oxygen from the lungs to the organs, as oxygen is necessary for tissue to be able optimally to use the sugar (in the form of glucose) that is also brought by blood. In normal conditions, cells can produce all the energy they need. It happens that oxygen is transported by attaching to the hemoglobin of red corpuscles (figure 3.5). Now, hemoglobin contains an atom of iron that happens to be magnetized. Pauling showed that depending on the various states of the hemoglobin, single electrons of its iron atom could change their configuration, seen in their radically different magnetic properties. Thus, although the molecule of hemoglobin is "diamagnetic" in the presence of oxygen ("oxyhemoglobin"), it becomes "paramagnetic" when iron has expelled the oxygen ("deoxyhemoglobin"). To clarify, this means that the oxygenated molecule of hemoglobin in a magnetic field presents completely negligible magneti-

[4]Linus Pauling and Marie Curie were the only people ever to have won two Nobel Prizes, one in Chemistry and the other for Peace in the case of Pauling, for his positions against atomic weapons testing.

zation, whereas the molecule of deoxyhemoglobin behaves like gadolinium and becomes a little magnet.

Normally, the hemoglobin that reaches the brain through the arteries is oxygenated, thus not magnetized. In the blood vessels of the brain, part of the oxygen is liberated and the red blood cells are enriched in deoxygneated hemoglobin. In the MRI scanner the red blood cells that reach the brain are progressively magnetized as they are freed of their oxygen. The blood in veins contains only 20 to 30 percent of deoxyhemoglobin, the magnetized form, but the march of these magnetized red blood cells changes the local magnetic field enough so that the water molecules in the immediate vicinity "feel" this change in the magnetic field, which is translated by a modification of their magnetization, an acceleration of their relaxation, and a slightly more rapid disappearance of the radio signal that they emit (figure 3.5). When Ogawa's rats breathed air lacking oxygen, the red blood cells reaching the brain already contained deoxygenated hemoglobin, which could not be completely replenished by oxygen in the lungs. When the quantity of red blood cells containing deoxyhemoglobin in the vessels of the brain increased, the effect of magnetic disturbance was intensified, and the water molecules completely lost their magnetization and no longer gave out a signal. This is why the blood vessels of the brains of Ogawa's rats darkened when they were in air lacking oxygen . . .

FROM RATS TO CATS . . . THEN TO HUMANS

The next question was to be expected: using this method would it be possible to follow in real time the variations in oxygenation of the blood in the brain while it is working? The English physicist, Robert Turner, in 1988 arrived at the National Institutes of Health (NIH[5]) in Bethesda, Maryland. Robert and I (already at the NIH for a year) quickly created a team because we had many common interests. He brought with him the revolutionary echo-planar imaging that was invented in 1997 by Peter Mansfield (which contributed to his winning the Nobel Prize), with whom he had worked. This method enabled us to obtain images of acceptable quality in a fraction of a second (instead of the usual minutes), but at the time remained very

[5] The prestigious institute of research in life sciences, endowed with a considerable budget.

difficult technically to put into place. In fact, this method requires altering the field gradients of the MRI scanner very rapidly and with very great intensity. With scanners intended for animals, smaller than those used for humans, it was a bit easier. Working with Turner, we thus sought to find out whether variations in the amount of oxygen in the blood could be detected in real time, a prelude to a possible application of the Ogawa principle to cerebral functional imaging. The animal chosen as our subject was a cat. We saw clearly that the "live" image indeed darkened each time the concentration of oxygen in the gas breathed in was reduced, and returned to its normal aspect when oxygen was restored (reference 3.6). The path was opening...

In the aftermath of this work, the competition was brutal: no fewer than four groups immediately began working to detect a change in oxygenation of the blood that might accompany cerebral activation: at Massachusetts General Hospital of Boston (where Belliveau worked); in Minneapolis, Minnesota (a collaboration with Seiji Ogawa); at the Medical College of Wisconsin; and us, at the NIH in Bethesda. In 1992, three articles appeared almost simultaneously, showing that it was effectively possible to follow in real time the variations of oxygenation of the blood in humans during stimulation: the back of the brains (where the primary visual fields are found) of volunteer subjects appeared whiter each time a visual stimulus (the famous blinking checkerboard) was presented and "faded" when it was no longer visible (figure 3.6; references 3.7, 3.8, 3.9). We could even show the images in the form of a film. This was the birth of functional brain imaging by MRI (often called "fMRI") as we know it today, and the method was then called BOLD (for blood oxygen level dependent), an apt acronym!

The Prowess of Functional MRI

And so we then had only to introduce a subject into the MRI scanner and obtain images continually while he or she carried out a task. There was nothing to inject, no radioactivity. The "tracer" enabling this feat is the hemoglobin of the blood that is oxygenated and deoxygenated as it travels between the lungs and the brain. An abundant tracer, naturally present and provided gratis by nature: what a difference from the gadolinium used by Belliveau and the radioactive water of PET! This conjunction of biology (neurovascular coupling), chemistry, and physics (the atomic and magnetic properties of the iron of hemoglobin and the proton of the hydrogen of water) is indeed "miraculous," as MRI enables us to observe how the magnetization of the protons of water in the field of the scanner evolves depending on cerebral activation and the variations in the blood flow that accompany it. Who would have believed such a thing just ten years earlier? This phenomenon did indeed already exist, however, as soon as the first subjects were placed in an MRI magnet, but no one had imagined it, and so no one had seen it: to be able to see something one must first know or believe it exists, or at least not have preconceived ideas.

Sure, the American scientist Keith R. Thulborn had shown in vitro that hemoglobin emitted a different signal on MRI images depending on its chemical state (oxygenated or not), but without making the connection with cerebral activity. In defense of the MRI physicists at the time, it remains that the effect was very small. Its revelation was possible on the one hand because Ogawa's experiments had opened our eyes, but also because the technology had matured: the acquisition of images in real time with the echo-planar method, computer software, and analysis issued from PET enabled the detection of changes captured in images.

The functional images obtained from the activated human brain are not all completely the same, and are not as contrasted as those observed for a rat deprived of oxygen. First, what one sees is not a darkening of the blood vessels in the activated regions, but on the contrary, a more diffuse lightening around the vessels in that region. This might appear paradoxical, but it happens that the activation of the brain notably increases the blood flow in the activated regions, which increases the amount of "fresh" blood, rich in oxygen, whereas the increase in consumption of oxygen from activity remains weak. As the quantity of oxyhemoglobin increases, the paramagnetic effect of the deoxyhemoglobin diminishes, and its disturbing effect on local magnetization of neighboring water molecules decreases. Recovering their magnetization, their signal rises (figure 3.6). The activated regions thus appear "whiter" than in a nonactivated state in which the vessels contain a certain quantity of deoxyhemoglobin. The effect is nonetheless very subtle and in general completely invisible if we simply look at the images directly. Here again, computer science plays a fundamental role.

THREE

Don't Think of Anything

With BOLD fMRI (as was the case, moreover, with PET), we obtain images sensitive to blood flow, but in a relative way. Irrigation is far from uniform in the brain. These "raw" images thus do not inform us of the state of absolute activation. In pathological cases they can show anomalies, which can be very useful. But for functional imaging in a normal subject, it's not enough. The key is to establish at least two conditions with the subject and to compare the resulting images. Where there has been change, we can deduce that there has been cerebral activity, by calling upon the hypothesis known as "insertion," according to which the brain works linearly, each new task being added to the underlying state. Thus, by comparing the images obtained when the subject is looking at an image (such as the checkerboard) with images of a "reference" condition, one hypothesizes that the only difference in the cerebral state is the presence of the visual stimulus. It is unfortunately quite obvious that many things can go through the head of the subject when he isn't "looking" at anything; he might, for example, be thinking of his coming vacation, even if we ask him explicitly to "empty his head," which isn't so simple, unless one is an expert in Zen Buddhism . . .

The insertion principle thus assumes that nothing changes in the cerebral state of the subject when the visual stimulus appears, except that which is directly linked to the stimulus. We can imagine the "rawness" of this principle, which is nonetheless used routinely in a great majority of studies in functional neuroimaging that are published today. If during the condition of reference the subject is thinking of the beach on his coming vacation, and that activates the visual cortex (we will see in the next chapter that this is the case for all mental visual imagination), the level of reference will clearly be affected. For the visual cortex a state of vigilance and a level of attention are also very important in the magnitude of the response to presented stimuli. The great difficulty in neurofunctional imaging is thus not so much obtaining images of the activated brain but correctly interpreting the images. The ball passes from the court of physicists to that of neuropsychologists, who must devise increasingly complex protocols of activation to take into consideration all the possible angles of interpretation, and to eliminate as much as possible any interference between the stimulation and the underlying mental state of the subject.

Proof through Statistics

But in the meantime, the ball has landed in the court of statisticians and image-processing specialists, who verify that the pixels described as "activated" by the image-processing software, due to a change in the signal between the activated condition and the condition of reference, indeed correspond a priori to changes in the blood and thus to a real activation. Indeed, the images contain a certain level of noise (whether it comes from little instabilities in the functioning of the MRI scanner or from spontaneous fluctuations in the activity of the brain). To limit this noise, we generally collect a series of images in conditions of activation and of reference, and the images are then averaged. Due to possible fluctuations in the time of the signaling of images, we alternate the conditions of activation and reference.

For example, we ask the subject to touch the tip of his thumb to those of his other fingers, in a sort of rapid piano-playing movement for 30 seconds, then to do nothing, then to do it again, and so on. (This task, apparently very simple, has nonetheless required the entire span of evolution for us to be able to do it with such dexterity, as no other animal can do it. A DNA sequence that evolved very rapidly in humans and not in primates has, moreover, been recently identified as being responsible for a human's ability to oppose the thumb to the other fingers on the hand.) During each 30-second period dozens of images covering all of the brain are taken and then averaged. In fact, the observed differences in blood flow are extremely weak, on the order of 1 percent. If the signal fluctuates spontaneously to a rate of 3 percent, one assumes that it will be difficult to conclude anything.

To make information appear out of noise is a challenge well known to astrophysicists, who must deal with extremely noisy signals coming out of space, which doesn't prevent them from obtaining stunning and reliable images of distant galaxies. In functional neuroimaging, specialists in image processing and statisticians are, like magicians, capable of drawing out of noisy images pixels that they confirm are activated. To do this they make subtle calculations that take into account the level of noise in the images and statistical fluctuations and their distribution in space and time, and also take into consideration the fact that the neighbors of an activated pixel are most likely activated, as well, or that an activated pixel will most likely be activated again on the following image (just as there is a greater chance that it will be nice the day after a sunny day—a weather prediction based on this principle would be better than chance!). In fact, just as meteorologists talk of a 70 percent chance of rain, statisticians only get partially wet. They don't tell us that a pixel, or a cerebral region, is activated, but that given the chosen parameters, there is a 5 percent, 1 percent, or even 1 in 1,000 possibility at least that a pixel has changed its signal by chance and is thus not in fact really activated. This degree of confidence is expressed on images using a color scale that enables neuropsychologists to evaluate the certainty of the different results of activation obtained.

The Homunculus Seen in fMRI

Regarding the movement of fingers, the motor cortex and the sensory cortex (related to sensation on the tips of the fingers) "light up" in the region that corresponds to the cerebral mapping of fingers (figure 3.7). The entire body is in fact represented in a very ordered way on the surface of our brain. The motor cortex where the neurons that command our muscles are found (neurons that may be over 1 meter in length and descend as far as tip of the spinal cord to connect with other neurons that go into the muscles) is located in the frontal lobe in front of the important fissure of Rolando (see figure 1.1). The other side of the fissure, in the parietal lobe, is where sensitive neurons that have traveled through the body after several relays terminate. Along this surface, toward the interior of the brain, there are neurons connected to the lower limbs and the feet, while those linked to the hand and to the upper limbs are spread at the top of the hemisphere. The neurons that command the muscles of the face and the mouth are on the external surface, at its lower extremity. In addition, the right side of the body is piloted by the neurons of the left hemisphere, and vice versa.

We can thus draw a map of our body on the surface of the motor cortex, which shows a sort of small human being called a "homunculus" (figure 3.7). The different parts of the body are represented with varying degrees of importance there—for example, the region of the feet is not very extensive compared to that of the hand because the movements we have to accomplish with our hands are more numerous and complex. Thus in the eleventh variation of Liszt's *Sixth Etude after Pagnanini*, the pianist's hands play 1,800 notes per minute, each of these movements being very precisely controlled.

This hyper-representation of the hand in the cortex, brought about by the development of an upright position that freed the hands, is of course unique to humans and has been refined throughout the evolution of the human species. It is what has enabled humans to acquire manual dexterity, and thus to make tools, to draw, and last to write. The Ancients already understood that relationship, for Anaxagoras of Clazomenae, of whom Plato was the disciple, wrote: "Man thinks because he has a hand."

The region corresponding to the face and the mouth is even more extensive, as it is connected to the very numerous possible movements in the face—facial expressions are extremely important means of communication—and with the movements of the buccopharyngeal area of spoken language. This development might also be linked to that of the hands: with the hands freed, it became possible to cook food, which softened its texture as compared to raw meat, and which then enabled jaws and the muscles that work them to develop into lighter structures, opening the airways that enabled language. But this of course remains to be confirmed . . .

THE TRAPS OF fMRI

Neurofunctional imaging has thus become a complex discipline to which many specialists—physicists, statisticians, computer scientists, neuropsychologists—contribute. Such interdisciplinarity is fundamental because individual participants are not always aware of the details of the entire picture, from the manipulation of the magnetization of protons, to statistical analysis, and the knowledge of cerebral states. Without this group effort the potential—and limitations—might be forgotten, and the interpretation of images distorted.

For example, the extremities of the temporal and frontal lobes (in the front of the brain) often show no activation in BOLD MRI images. It would, however, be a serious error to conclude that nothing is happening in those regions (and *a fortiori* that they can be removed by a neurosurgeon during the excision of a lesion without any resulting problems!). Functional images, with their scale of confidence in color, are obtained most often with the rapid echo-planar method. Such images, however, have proven to be of mediocre quality in representing anatomy, and are much more subject than standard images to distortions that may be caused, as we have seen, by the sinus cavities of the face (see figure 2.1). And so images of "activation" are often incapable of showing activations at the tip of the frontal and temporal extremities of the brain, whereas the "anatomical" images upon which they are projected to better recognize the anatomy are of much better quality and show the entire brain: the tips of the frontal and temporal lobes are completely visible here, whereas an activation cannot be detected.

Generally, it is very difficult to determine the absence of activation when a given region is not "lighted up" in images following stimulation. Apart from defects in the base images, any increase in noise or fluctuation of the signal can mask small changes in local magnetization provoked by an influx of oxygenated blood. Registration algorithms can to a certain degree correct the effects of head movements—for example, by utilizing certain hypotheses on movement—but such corrections become ineffective if the movement is too great. We can imagine, then, that in the current state of the technology, functional imaging encounters even greater difficulties in noncooperative subjects, such as some elderly or agitated patients, or children. The best subjects are in fact the experimenters themselves, because they understand better than anyone what is at stake with the problem of extraneous movements, and because they are very motivated to obtain good results.

The Missing "L"

As regards challenges with fMRI, a problem that quickly arose involves language. It is in fact difficult to make subjects speak inside the magnet because of the effects of movement. Similarly, the great noise of the MRI scanner can render practically inaudible the auditory stimuli pertinent for language (we can do this today with certain limitations). We thus arrived very quickly at the following idea: instead of having subjects speak inside the scanner, why don't we simply ask them to *think* of words, in their head, silently? For example, to think of the names of animals, or of words beginning with the letter "P" for 30 seconds, then of nothing, then of the names of clothing or words beginning with "R," and so on.

Would the corresponding cerebral activation be sufficient to be detected by MRI? We checked. The results were immediately positive, as the Broca area, behind the frontal lobe of the left hemisphere, in particular but not exclusively, "lit up" during each period of silent evocation of words, and was extinguished during the periods of rest (figure 3.8). At the beginning of this project, one of my colleagues played a trick on me. Whereas I had asked him three times to think silently of a series of words beginning with "P," then "L," then "R," each for 30 seconds, I could observe a response for P and R, but not for L. The team began to get discouraged: what if the method was not as reliable as all that, after all? Crestfallen, we removed our colleague from the scanner, ready to explain to him that the experiment hadn't worked very well . . . He asked us if we had seen activation for the three letters, and we admitted that we had seen nothing for "L." In fact he had thought of nothing during the 30 seconds of "L"; he had simply wanted to test our scientific acuity! The experiment had thus worked beyond our hopes.

The mere fact of thinking is enough to cause a small increase in blood flow and oxygenation of the blood in an activated region, thus modifying the quantum magnetic state of the protons of water molecules near the small blood vessels, and thus the MRI signal. This change is minute, of course, but it is detectable with the appropriate statistical tools. We are dealing, in fact, with nothing less than the (physical) action of the mind over matter, revealed by MRI! With functional MRI it indeed becomes possible to see the brain think . . . BOLD MRI is constantly improving with the use of more sensitive and effective methods of acquiring and processing images. Instead of lengthy stimulation in blocks of time, for which the brains of subjects are assumed to remain in a constant state and to carry out tasks with the same intensity, it has become possible to record images corresponding to single activations, such as a simple click of a mouse, for example. Functional MRI has become "event related." But it is now high time finally to discover what the images of our thinking brain reveal.

FOUR

THE MAGNETIC BRAIN IN ACTION

Functional MRI (fMRI) enables us to see the brain in action, its images show us which regions of the brain are activated when we carry out a given task, such as looking at a picture or moving our fingers. These simple tasks were very useful, especially in the beginning, in demonstrating that it had become possible and relatively simple to see the brain think with fMRI (many researchers were skeptical). But neuroscientists, neuropsychologists, and even neurologists wanted more than simple tasks; they wanted to put their complex paradigms to the test with this new technology.

And so in 1992—fMRI was only one year old—I began to study *mental* imaging, showing that fMRI allowed us to see activation of visual regions using both real and imaginary images (reference 4.1). Looking at a (real) image activates first the primary visual cortex located around the "calcarine fissure," in the back of the brain. This is where the visual paths converge, from the retina via the optic nerves that serve as relays in two small structures at the base of the brain, the "lateral geniculate body," out of which comes a very large network of white matter fibers, the "optic radiations" (see figure 3.4). The image that reaches the primary visual cortex is in fact immediately dissected, decomposed, analyzed (for example, for its content in color, the orientation in space of its constitutive elements, its movement, and so on) before being sent back, completely deformed and in some way predigested, to the front of the brain. It is only then that we realize that the image in front of us is that of our . . . grandmother. Some neuroscientists think there are neurons that are very specialized in the detection of objects or people (the neurons of the aforementioned "grandmother"), some cerebral regions detecting specifically animals, others human faces, and others, words.

A Cat in the Brain

But what takes place in our head when we "see" an imaginary cat, with our eyes closed? Where is it? Insofar as there are Siamese, Persians, and alley cats, or white ones and striped ones, nothing indicates a priori that a term as generic as "cat" will always activate the same regions, and in particular the primary visual region that is connected to the real world. No one knew the answer to this question and it was the object of controversy, many thinking that the primary sensory regions dealt only with the real world.

We asked a person lying in the MRI scanner to think of a cat (figure 4.1). Eight seconds later a signal indicated to him that he should stop thinking of the cat (or of any mental vision). And 14 seconds later, in order to have him return to the initial image, we asked him something like: "Does the cat have pointed ears?" The fMRI images showed the regions involved, notably the region of the calcarine fissure with the primary visual cortex, activated then by the imaginary vision of the cat, as if the cat had really been seen (figure 4.1)! This region was activated in every subject in the experiment, regardless of the animal imagined, from an ant to an elephant. Thus the simple fact of thinking, of creating a mental visual image, activates the primary visual cortex (reference 4.2). The proof was there: the visual cortex is thus not so "primary" after all, and is not solely connected to the real visual world. But this experiment also revealed the incredible potential of fMRI: the fact of thinking alone, of imagining, is enough to change the magnetization of water (via the change in oxygenation of the hemoglobin of the blood in the vessels irrigating this cerebral region), which the MRI scanner shows us. It is thus indeed our thoughts that this technology reveals, or at least (for the moment), the cerebral networks involved in the thought process. The activation visible through MRI is, moreover, even stronger when we ask the subject to recall the shape of the animal's ears; he must then search and "zoom in" on his memory to recover that detail.

The "primary" regions of the brain thus connect us not only to the real world that surrounds us, but also to our mental world. We then verified that this is true for the primary motor and sensory regions, as well. The subject, instead of tapping his fingers against his thumb, simply had to imagine the movements of his fingers, and the sensation of the tips of his fingers touching each other. To show that there wasn't any "cheating" we also recorded the electrical signals in his arms to indeed verify that no muscle was contracting. The fMRI images were clear: here, too, simply imagining the movement activated the primary motor region that commands the fingers, as well as the opposing primary sensory region (see figure 3.7).

These results completely validated "mental training" methods. High-level athletes "imagine" their movements in advance, a golf swing, or a succession of markers in a ski run, which increases their performance by preparing their cortex for the chain of elementary movements. What is more, a group of researchers has shown that the corresponding muscles were developed through this practice, which is a sort of "body building" through thought. I experimented with this approach myself with the piano—not moving it, but playing it. Since I travel often for conferences, I can't always find a piano available, and I can only imagine that I'm practicing pieces off the score . . . When I return home, I often note that I have indeed made progress when I play the pieces I had mentally practiced for a few days. On a completely different level, Vladimir Horowitz and Arthur Rubinstein revealed that they, too, used this method to improve their performances. And it is no doubt this possibility that enabled the Chinese pianist Zhu-Xiao Mei, who plays Bach's Goldberg variations so wonderfully, to preserve, develop, and master her art during the years she was imprisoned in the Maoist reeducation camps severely lacking in pianos.

A unique quality of the primary visual cortex is that it is "retinotopic": the (real) world is projected on it upside down, left to right, and vice versa, like on an electronic light sensor in a camera. Granted, the image is very deformed, and its center is hypertrophied (which enables us to distinguish details, to read), but the correspondence exists. Thus, when we look at a horizontal object, a pen for example, only part of the visual cortex is activated (the horizontal "meridian"), whereas a vertical object activates the vertical meridians (there is one on each side of the visual cortex). Our fMRI images show that one only has to think of a vertical or horizontal object to activate the associated meridians. In other words, studying fMRI images enables us to guess the thoughts of the subject and determine, with a high rate of success, whether he is thinking of a horizontal or a vertical object (figure 4.2)! (But the kind of pen and the color of the ink are still unknown.) Building on this ability, Bertrand Thirion in our laboratory has been able, using a computer algorithm, to decode the fMRI images of the visual cortex of volunteers imagining letters of the alphabet, letters that we then identified fairly successfully (reference 4.3). Sure, we first had to "decipher" the code of these subjects—that is, establish for each of them the correspondence between the real visual space and their visual cortex by means of calibrated stimuli—and thus we weren't really reading their thoughts, but we were getting amazingly close . . . What was really happening in this visual cortex?

MENTAL READING

In his *Letter on the Blind for the Use of Those Who See* (1749), Denis Diderot explores visual perception and wonders what a person blind at birth who, following some kind of operation, recovers his sight actually sees. Relying on the sense of touch fascinates him, since it allows the blind person to construct another perception. "He has his soul on the tip of his fingers," writes Diderot, which indeed raises the question of the location of the soul.[1] Even if he was unaware that one day we would see the brain in action, Diderot asked the right question: do the blind use the same regions of the brain as those who see? To locate the soul through MRI, however, we will still have to wait a bit.

One of my Japanese colleagues, Norihiro Sadato, who perhaps read Diderot, asked blind subjects to read Braille script (with their index finger, as is customary). He noted that their primary visual regions are activated by reading Braille (reference 4.4)! A blind person thus still has functional visual regions and he "sees" in some way with his fingers. We find ourselves with the ambiguity of cerebral function: what is "seeing," and does a blind person "see?" Granted, we might think that nature takes care of things and the brain has "recycled" the visual cortex, which became useless, to turn it into something else. I find this hypothesis based on long-term cerebral plasticity a bit reductionist. Another hypothesis suggests that one or rather several precise mechanisms, certainly programmed genetically, intervene in the deepest part

[1] Diderot's letter, considered to be dangerously atheist, earned him three months in prison.

of the visual cortex, giving it its functional specificity, which is not exclusive to vision. In a certain sense, the fact that the blind are deprived of sight reveals the work that the visual cortex does naturally through sensory data processed through paths other than vision. This is suggested by the recent work of a team in Boston. After only five days of an intensive study of Braille, seeing subjects, but with their eyes blindfolded, began to recognize the script tactilely, and this was not the case with seeing subjects not wearing a blindfold. fMRI revealed that there was a response in the visual cortex during tactile stimulation, but only in the "blind" subjects. This response disappeared 24 hours after the blindfold was taken off (reference 4.5).

Even more convincing was the temporary disturbance of the functioning of the occipital lobe (where the visual cortex is found) through transcranial magnetic stimulation (TMS, which consists of subjecting part of the brain to a very intense pulsed magnetic field produced by a coil placed next to the head for a very short time). TMS prevents the subjects from reading Braille correctly, which suggests that the visual cortex indeed has an active, natural role in the processing of information of tactile origin (reference 4.5). The underlying mechanism could be the spatial "binding," the virtual connection of points in space to bring out traits or shapes, a necessary stage of the information that passes through it before a more global processing by other cerebral regions. This process is obviously essential to vision, but also for visual mental imaging (the pointed ears of the mental cat!) and of course in recognizing the patterns of the dots that make up Braille. The "visual" cortex would thus be more of a "metric" or spatial cortex.

Recent work also suggests that people blind from birth who are trained to construct a sort of visual mental image from sonorous stimuli (somewhat the way bats orient themselves using ultrasonic signals that they emit and which are reflected off walls) activate their visual cortex for this task. We can thus imagine that a given cerebral region, going from common, elementary tasks unique to that region, might carry out diverse tasks on a more "macroscopic" level in function of the data it receives, ranking the data preferentially. We have recently discovered that the primary visual regions of those blind at birth—in particular, the regions on the left side—are also involved in the understanding of spoken language, especially for the understanding of words and phrases (reference 4.6). By contrast, these results highlight that the cerebral regions usually associated with language are not necessarily the only ones capable of such a function. Indeed, and fMRI is there to remind us, the brain is a very complicated organ.

The implications of the results obtained through fMRI are very great, in that they allow us to imagine ways to repair certain deficiencies in our brain and to improve the quality of our lives, and in particular those of patients afflicted with neurological illnesses or handicaps. You no doubt know the little game called "rock, paper, scissors," which consists of showing your hand in three different positions—as "paper," as "rock," or as "scissors"—one of the choices always winning over another. A team of Japanese neuroscientists conceived of a variation of this for fMRI. The MRI imager indeed enables us to identify and differentiate in real time the subregions of the motor cortex involved in these three hand positions. If a computer automatically analyzes these images and sends electrical signals to miniature motors of an artificial hand, located at a distance, the hand is animated, in real time, by the same movements as those of the hand of the person in the MRI scanner (figure 4.3). If the person just imagines the movements, without doing them, the artificial hand still carries them out, the machine being directly piloted by thought! Using this principle Toyota conceived of a wheelchair for the handicapped piloted by thought, the prelude to a "mental" remote control replacing vocal commands (which itself has replaced manual controls). Its paralyzed user thinks (in Japanese, of course) "advance," "to the left," "stop," and the chair carries out the command . . . We no longer use MRI for this (too cumbersome for a wheelchair), rather electrical waves emitted by the brain and received by electrodes placed in a helmet.

The prospects of the technology are even more encouraging because we don't need to use signals emitted from the cerebral region corresponding to the function to interface with a machine. This is important because the cause of the anomaly (paralysis, for example) is often in the same region that is damaged. Let's look at another game: virtual ping-pong. In normal conditions this involves making a paddle move on a computer screen using a joystick to hit the ball that is moving on the screen. Two people can play. In our setup, the two players are both lying in an MRI scanner (you need two, and the experiment is a little bit expensive) and cannot see each other. Nor do they have joysticks. The only thing they see is the ball and the position of the paddles on the screen placed in each scanner. Their mission: to invent a criterion of their choice whose intensity can be controlled through thought.

This can be imagining a more or less intense red, or a more or less loud sound. During that time, the images are captured and analyzed by a computer. It will quickly pinpoint a cerebral region (or several) that is activated with variable intensity, and it will "learn" to use the signal from that region to pilot the position of the paddle on the screen. After a few minutes, the learning is done and the two subjects begin to play "brain pong," the position of the paddles being moved by their thought, via MRI and the computer, but—a remarkable fact—without their motor cortexes being involved! One can imagine the potential of this discovery, of the possibility for cerebral "reprogramming" via the use of any cerebral region to carry out tasks absolutely unforeseen by our genes—in particular, if a region of our brain has been destroyed through illness or an accident.

"When Things Are Bad, Look at Yourself in a Mirror"

The French translation of this Chinese proverb is "quand tout va mal regardes toi dans le miroir." It is even possible to reprogram one's own brain in *real time*, and fMRI is beginning to be used successfully with patients in what we call "biofeedback." For example, volunteers were subjected to a nociceptive stimulus (a sensation of burning on the left hand with a 48°C probe) for periods of 30 seconds (reference 4.7). In real time they were shown the image of a flame whose intensity reflected the level of activation of the rostral cingulate anterior cortex of their brain (figure 4.4), a region known to perceive and regulate pain. Their mission was to "control their mind" to increase or decrease the height of the flame. After a few minutes the subjects were able to modulate the flame through thought. What is remarkable is that their sensation of pain corresponded to the variations of the flame height, either increasing or decreasing. Other subjects in a similar test, but without MRI and without the feedback it enabled, were incapable of modulating the sensation. The experiment of biofeedback through MRI was then carried out again with patients suffering from chronic pain, without nociceptive stimuli this time, as these patients had unfortunately been suffering naturally and continually for many years. They were considered therapeutic failures, since no medication or psychological treatment had been able to help them. Here, again, they were able to modulate the activity in the rostral cingulate anterior cortex of their brain, via the image of the flame they were shown, and thus to control the sensation of pain, especially if the training session in the MRI scanner was long. What is more, the ability to control the level of pain continued after the MRI session, as the patients explained that their pain had dropped in intensity without their necessarily having had to pay attention to it. These encouraging results, however, remain very preliminary and are not used yet in medical practice.

A similar study was conducted with chronically depressed patients who were resistant to any treatment. For them, the feedback consisted of having them modulate in real time the activity of a region of their brain involved in pleasure, their mission being to increase, through thought, the activity of that region. Here again, the results were remarkable; some patients rid themselves completely of their depression after the fMRI exam, and thus no longer took any medication. Positive results were also obtained in Parkinson's disease, with patients learning how to control their brain, via fMRI, to improve their motor performance. There still remains a great deal to be done to better understand this biofeedback method, its long-term effectiveness, and to identify the patients who might benefit from it (it is certainly not for everyone). This approach has not yet been entirely validated, but it has enormous potential, as fMRI is becoming a sort of intrusive mirror from which one can hide nothing, in which the patient observes his own brain directly, and no longer through an intermediary, through the words of a psychiatrist, for example. It is still too early, however, to know whether the bed of the MRI scanner will one day replace the sofa in a psychiatrist's office. Direct communication between individuals, both visual and verbal, remains a fundamental principle of human social behavior, which unfortunately tends to be forgotten with certain so-called modern modes of communication, such as e-mail, and the term "social network" that is given to certain electronic means of group communication is perhaps not the most appropriate term. The word "virtual" has perhaps been a bit too quickly forgotten, whereas our world (and our brain) remains quite real, as we are shown, alas, every day in the news.

But let's return to our fingers, upon which we can always count, and to the motor cortex. fMRI has indeed revealed the cerebral regions involved in calculation, but this is not what I want to stress. Certain populations have learned to count in ways other than on their fingers: Asians have long used abaci (now they are usually found in secondhand shops rather than schools, as calculators have taken over) and count . . . on their beads! In any case, this is what fMRI shows. Those abaci have so infused the lives of children, at an age when cerebral plasticity is at its greatest, that they mentally use their abacus in mental calculation, which moreover often makes them more proficient than we are. As they mentally manipulate the beads, it is thus not surprising that fMRI reveals the activation of their motor cortex in operations of mental calculation. Although they count as we do "with" their fingers, but, while mentally imagining the movements on their abacus, they do so much more efficiently than they would "on" western fingers, which are limited to ten.

Similarly, we often speak "with our hands" (some more than others), and the Japanese even speak with their bodies, bowing often as a polite accompaniment to what they are saying to their interlocutor, even if the other is only on the phone . . . But they sometimes also speak with their fingers. The sounds of the Japanese language can be ambiguous, as can the kanjis, "Chinese" characters used to represent them. Japanese schoolchildren spend many years learning to write these kanjis, some of which are made up of twenty-three strokes of a pencil that must be made in a very precise order. Some kanjis have at least two pronunciations, and sometimes up to seven, and we can understand that traces of them remain in a learner's brain. Thus, when a Japanese adult speaks or listens, it isn't uncommon to see him move his index finger as if he were writing, a sign of the mental effort involved in the motor cortex, as revealed through fMRI, to be very sure he is identifying the corresponding kanji in case of doubt: thus the sound perceived by the auditory cortex in the brain is converted to a visual mental image (the kanji) via a mental representation by the hand.

SINGING IN THE BRAIN

fMRI shows us that the pianist does exactly the opposite, translating the score that he deciphers mentally via the virtual movements of his fingers into sounds that his brain alone perceives. Some have music in their skin, but pianists clearly have music in their fingers. A study carried out with twelve piano students at the University of Cologne in Germany, who practiced on average 22 hours per week, showed that when they looked at the score of *Triolak* (from Bartok's *Mikrokosmos*), the same motor and auditory regions were activated as they would be during a real performance, but less intensely (reference 4.8). What is more, when a pianist was shown a video recording of the fingers of another pianist playing on a keyboard, with the sound off, the (professional) pianist was able to distinguish whether it was a piece of music or a random striking of the keys (which a nonpianist can't do), and even recognize the piece being played, if he was familiar with it (even without necessarily having played it himself beforehand) (figure 4.5; reference 4.9).

This coding between the visual world, perceived through the occipital lobe, and the auditory world (virtual here) re-created by the temporal lobe, passes through the left "planum temporale," a region that is very involved in music. Many cerebral regions in fact participate in the development of movement: "the most intricately and perfectly co-ordinated of all voluntary movements in the animal kingdom are those of the human hand and fingers, and perhaps in no other human activity do memory, complex integration, and muscular co-ordination surpass the achievements of the skilled pianist . . . ," writes Homer W. Smith (*From Fish to Philosopher: The Story of Our Internal Environment*, 1959, p. 205). Playing Bach's *Concerto in F* thus reveals, beyond the activation of the motor cortex controlling the movement of the fingers, the activation of the supplementary right motor region, involved in programming the coordinated movement of the two hands (reference 3.3). The cerebellum, the small organ located beneath the brain, is also activated in the back, in connection with the processing of auditory information, and in the front, to smooth the movement of the fingers and the hands and make them harmonious. But certain regions can also be deactivated—in particular, those involved in attention and concentration (which we don't see in the practicing of scales)—perhaps when the musician is "carried away" while playing (inspiration!), as if the mind of the composer were channeled through him, taking possession of his brain. This situation is generally accompanied by a deep silence in the concert hall, followed at the end of the piece by thundering applause. I, myself, have experienced this feeling of becoming one with the music and the instrument, a sense of being elsewhere, my feet no longer on the ground (but still in contact with the pedals). A famous actor in Japanese Nô theater once told me he had had similar experiences. Nô theater is known for its slowness and its apparent lack of action, the actors' movements being extremely measured. In fact, the actors (who often incarnate demons) are literally possessed by their role, their heartbeats increasing to a level worthy of a sprinter at the end of his race.

Musicians of course also have music "in their head," and they do not always need real sounds to feel the music they write or that they interpret from the score. This is obvious for conductors who must often work in a "without orchestra" mode (not to be confused with *karaoke*, which does mean "without orchestra" in Japanese). Of course, they are also more sensitive to real sounds than others are. The auditory cortex is more broadly activated in pianists than in nonmusicians. But real sounds, although not heard by the ear, can also be "heard" by the brain of musicians or other experts, such as sound engineers. We know that the human ear doesn't hear beyond 20,000 Hz (this threshold diminishes with age), and the systems of encoding/compaction that we find on our electronic gadgets largely take this into account. However, Balinese gamelan music is rich in sounds beyond 20,000 Hz. fMRI shows that some nuclei of the brainstem are sensitive to these sounds, which are no doubt transmitted mechanically by boney structures, which Sony must have taken into account in developing its "superaudio" encoding technology. Lower in the brain, melodies that are sung, especially when they are known, activate the left side of the brain (but they usually also have words), whereas the right brain deals with the sound of instruments. There are many interactions between regions of language and music, but there are also differences, and their respective networks cannot be superimposed. Moreover, cerebral damage may alter a perception of music, but not of language, and vice versa. The American president Franklin Roosevelt and the Cuban revolutionary Ernesto "Che" Guevara suffered from amusia, a disorder that prevents the processing of the music they heard. Similarly, a French music-loving mathematician who became "amusical" said he knew when the *Marseillaise* (the French national anthem) was being played when he saw everyone standing up. At the end of his life Maurice Ravel, who had developed nonidentified lesions on the left hemisphere of his brain, had trouble composing (in 1933 he could no longer write or swim). However, he heard music: "I will never write my *Jeanne d'Arc*, my opera is in my head, I can hear it, but I can no longer write music. I have lots of ideas but they disappear when I try to write them," he confided. The illness had begun some dozen years before his death in 1937, and some have suggested that his music from that time perhaps reflected his illness. The eighteen repetitions of the theme from *Bolero* (the most played piece of music in the world, played on average every 2 minutes) might correspond to a syndrome of perseveration that we observe in afflictions of the frontal lobe. The complexity of the score, with its superimposed binary and ternary rhythms, the orchestration with twenty-five different tones, reflect a style different from his other, earlier compositions, perhaps related to a greater compensatory participation of the right hemisphere in his creative activity. All of this, of course, remains a conjecture that, alas (or for the best?), will never be proven.

WHAT SIDE DO YOU SPEAK ON?

Looking back at our hands and fingers, we all know that we are asymmetrical, and that one half of our body dominates the other. Most of us are right-handed, some are left-handed, others are ambidextrous. This is a known fact and in general poses no problem. This was not always the case, however, and until recently left-handed children were often "pressured," forced to write with the right hand in school. Moreover, the French language is filled with expressions (wrongly) suggesting that "rightness" is good (one studies *Droit* [law], not *Gauche* [left]!), but that it is unseemly to be left-handed ("elle est mal-a-*droite*," [clumsy], "il est *gauche*" [awkward], and so on), and this has been going on for a long time: the French word *sinistre* [sinister](negative) comes from the Latin *sinister* which means "on the left," whereas *dextérité* [dexterity](positive) comes from the Latin *dexter*, "on the right" . . . There are so many innuendoes in these apparently harmless words, but they are entirely linked to the functioning of our brains!

For it is of course our brain that is responsible for this asymmetry between the left and the right, since it is itself functionally asymmetrical, as we have known since the very first pages of this book. Let's note that nature has counted on "cohabitation": the right side of our body (in general) dominates, but its commands are piloted by the left hemisphere of our brain. Speaking about left and right sides, whether in the context of politics or brain function, reminds me of a famous, but perhaps not so smart, U.S. politician who had an MRI scan. His doctor reported to him that there seemed nothing right in the left side, and nothing left in the right side . . . Why are the regions of spoken, understood, or written language generally on the left side? We do not yet have a clear answer to this question. What we know, however, is that this lateralization appears very early in our life, since it is already present in 2- to 4-month-old babies, long before they speak (reference 4.10). The study of babies of this age through fMRI is not very easy, but the team of my colleague Ghislaine Dehaene was able to overcome this obstacle: keeping the babies immobile in the scanner while they are awake. It takes a lot of patience to wait for the right moment, and especially to know when to give up when the necessary conditions are not being met. The success rate is thus not very high, but we have learned that when a 2-month-old baby hears the (recorded) voice of his mother speaking to him, the left hemisphere, in the temporal lobe, where an adult responds to significant stimuli on the linguistic level, is indeed activated (figure 4.6). By contrast, other sounds, without any connection to language, provoke a bilateral auditory response. However, this baby is still far from speaking. Does he understand what is being said to him? That is difficult to know. And yet his reaction is not the same when he listens to a recording of his mother's voice played forward and then in reverse. There is activation in the anterior part of the right frontal lobe when the recording is played forward, whereas no response is visible when the recording is played in reverse, although the recording has the same frequency characteristics in both directions (figure 4.6). Babies thus know the difference between what has linguistic sense and what doesn't. When in doubt, we should be careful of what we say around them . . .

But only around 85 percent of the population has language lateralized on the left. The others either have language predominately on the right (often left-handed people, but not always), or divided equally between the two hemispheres (some left-handed and ambidextrous people). Women are also generally a bit less lateralized than men, with language being more bilateral in them. Granted, we don't really need to know how we are cerebrally lateralized for language in order to live—except in particular cases. One of these cases involves epileptic patients, for example, who must undergo a surgical procedure. Epilepsy is most often well controlled by medication, but some uncontrollable forms of epilepsy resist treatment. These patients can have very frequent epileptic seizures, which makes their lives difficult, especially if they are school-aged children. In those cases, we want to pinpoint the cerebral region at the origin of the seizures, and remove it from the rest of the brain. If the region is close to those linked to language activity, there is a risk that the patient will become aphasic after the neurosurgical procedure. Thus, before deciding to operate it is imperative to know whether the lesion and the language regions are on the same side, and if that is the case, at what distance from each other. Hand dominance, right or left, is not a reliable indicator, as we have seen. The test that is used as a point of reference is called the "Wada" test. It consists of injecting through the carotid artery, on the left and then the right, an anesthetic agent believed to put the cerebral hemisphere on the side of the injection to sleep. The patient is then asked to speak, to read (for a child, to name the images he is shown): if he is no longer able to, that means the anesthetized side is that of language. Besides the rather invasive and somewhat risky nature of the procedure, its reliability is not entirely foolproof. There is, in fact, a little "communicating" artery that links the vascularization of the two cerebral hemispheres. If the anesthetizing agent rushes into this artery and the side that is anesthetized is not the one we think, there can be dire consequences, as you can imagine.

My NIH colleague Dr. Lucie Hertz-Pannier and I were the first to propose fMRI for children to determine the lateralization of language before a surgical intervention. These children had only to lie down in an MRI scanner and silently think of words (the names of toys, food, clothes, and so on.) The images obtained then quantitatively showed which side of the brain was the most activated, using a previously defined index of lateralization (see figure 3.8; reference 4.11). No more injections, no more doubt about the localization; we were able to see directly the activation of the brain on the side of language. The experiment was completely successful, and it corresponded perfectly with the Wada test (which was still being used). Back in France, we of course continued this work, determined to remove the Wada test from the list of preop exams. Our first patient was a little 6-year-old boy suffering with Rasmussen's encephalitis. This chronic inflammation of the brain of unknown origin that occurs in children literally consumes the brain and causes it to atrophy. The epileptogenic lesion was on the left. The Wada test suggested that language was on the right. But the fMRI showed that the language regions were in fact on the left! The first case of disagreement for us . . . What to do? The neurosurgeon hesitated for two years, but finally, faced with an increase in the number of seizures, decided to operate. The little boy lost his speech. The fMRI was thus correct, but we couldn't gloat. And yet, after 2 or 3 years the little boy started talking again, almost normally. His parents gave us permission to look at his brain again with fMRI: the region on the left that had been operated on was very visible, and showed no activation. But the areas of language had clearly passed to the right, thus confirming the extreme plasticity of the brain, at least in children (figure 4.7; reference 4.12). This is why neurosurgeons are always faced with a terrible dilemma before very young epileptic children who must be operated on: should they intervene as soon as possible to take advantage of the plasticity of the brain in the event of complications, or wait, in the hope that the epilepsy might diminish spontaneously.

Today, fMRI competes with the Wada test in presurgical considerations, both among adults and children (in addition, cerebral activity in the affected region is verified directly with a series of electrodes that are placed on the skull during the operation) and is used for other diagnostic purposes in addition to determining the lateralization of language. For example, if a cancerous brain tumor must be removed surgically, it is important for the neurosurgeon to know whether removing surrounding regions out of precaution will have functional consequences for the patient. Thus, if the tumor is adjacent to the motor cortex, the patient might become hemiplegic, and the intervention must therefore be avoided, lowering the patient's chances of survival. If the tumor touches another, more central motor region, the "supplementary motor region," the patient will also be paralyzed following the intervention, but that paralysis will completely disappear after about 6 weeks (figure 4.8; reference 4.13). With fMRI we can functionally localize very precisely the primary motor region and the supplementary motor region. If the latter might be affected, or removed during the procedure, the neurosurgeon can safely intervene, after warning the patient that he will be temporarily paralyzed following the operation. Thanks to fMRI, postop functional complications have been reduced, as has the duration of hospital stays following an operation, a significant improvement on medical and economic levels.

An affliction of the cerebral regions connected to language can, however, reveal surprises. We have all heard those strange stories, sometimes called miracles, of patients who, following an accident or an illness, lose the ability to speak their native language, but begin to mutter in a foreign language. fMRI is not yet a method recognized by the Vatican to identify something as a miracle, but it can shed some light on the Broca area and bilingualism. Let's take the case of truly bilingual people, those who most often emigrate to a foreign country shortly after birth, emigrants to the United States, for example, who as adults speak fluently both the language of their adopted country and the one their parents continue to speak at home. When we ask these people to think of words in one language, then the other, fMRI images clearly show activation of the Broca area, and its localization in the inferior part of the left frontal lobe is the same regardless of the language. If the experiment is then conducted with people who, like most of us, have learned another language at around 11 years old, we see that there are two Broca areas (or a split Broca area) that are certainly very close together, but are different, nonetheless, with one for each language (figure 4.9; reference 4.14). We can thus understand that if one of those areas is damaged by accident or illness (which is very rare), the patient has no other choice than to speak in the language he has left, either his native language (then the damage goes unnoticed) or the foreign language that he has learned more or less well (and then it is called a miracle). But when this work was published and reported in the media, many interpreters became nervous. Would they henceforth be tested via fMRI before being hired in order to find out if they had one or two Broca areas, and thus if they were really bilingual? This is where fMRI encounters limitations, limitations that unfortunately are sometimes ignored by those who make images say a bit more than they actually say. Whether a person has one or two Broca areas has nothing to do with his or her ability to speak a second language. One can be quite proficient in a foreign language and have learned it only at age 11, even if we now know, given cerebral plasticity in children, that it is better to learn a foreign language very young (but then is it truly "foreign?"). What fMRI shows above all in subjects with two Broca areas is that different circuits were used in learning the two languages, on the one hand because the brain "grew" between the two processes, and distinct cerebral networks were recruited for each, and on the other hand, because the learning methods, total immersion or a scholarly approach, were different. The interpretation of fMRI images thus remains a difficult art.

An Expensive Lie Detector

A perfect example of this difficulty is that of lie detection. Today, there are two Internet sites of U.S. companies, one based in California and the other in Massachusetts, who are marketing a lie detector via MRI. The idea seems to have come from a study published a few years ago showing that the cerebral regions involved in the indication of "true" and "false" memories are not the same. From that to placing someone in a scanner and having him relive his memories to find out whether he is telling the truth took only one step, and it was taken. fMRI has been proposed within the framework of several legal actions in the United States (sexual harassment, murder and rape, and so on), but it has still never been accepted on the judicial level. This is not the case in India, where on several occasions fMRI has been used in court, including in at least one case of homicide leading to the death penalty. The Indian Supreme Court has nonetheless recently rendered the use of fMRI to detect lies unconstitutional.

Indeed, we must remain very careful. Published scientific studies are carried out under extremely controlled conditions in which the subjects are by definition volunteers who take the studies very seriously (especially if they are themselves the authors of the study!). Often the results are inconclusive and the conclusions are drawn from an averaged combination of several images from several subjects, without any significant results being obtained for a single subject. Even when results are obtained individually for each subject, we are working in laboratory conditions that are quite different from those in the real world.

However, fMRI can also have positive effects on the judicial system. Several studies have in fact been devoted to the brains of adolescents, and to their sometimes risky behavior, a societal phenomenon with important health consequences (alcoholism, drugs, unprotected sex, and so on). In one of these studies, the photo of a smiling or a neutral face is shown to an adolescent on a screen for half a second. The photo of a second face follows a few seconds later and the subject must push a button if the second face is of the same type (smiling or neutral). Everything is repeated (inside the MRI scanner) a dozen times. Compared with adults or children, adolescents are wrong most often with smiling faces than with neutral faces, which indicates a loss of impulse control vis-à-vis attractive stimuli (reference 4.15). This counterperformance is correlated with hyperactivity in a small region of the frontal lobe ("prefrontal cortex ventral striatum") suggesting in adolescents a hyperrepresentation of attractive stimuli in that region. Thus, on the basis of such fMRI studies showing that the frontal cerebral regions involved in decision-making are immature in adolescents, the death penalty in some states in the United States can no longer be applied to criminal adolescents under 16, even after they become adults, and the punishments reflect lesser guilt for juvenile conduct. fMRI has thus clearly opened the door to law enforcement and the judicial system, and this is only a first step.

Although fMRI doesn't have the capacity to force a presumed guilty person to talk, it has the means to make him listen and his brain speak, possibly against his will. Our colleague Christophe Pallier performed fMRI studies on a series of subjects who listened to bits of text read in different languages, some known by the subjects, others not (Polish, Hungarian, and so on). The left temporal-parietal region was activated only when the language was understood, and otherwise remained inactive (reference 4.16). The results did not require the cooperation of the subject, who remained completely passive, his brain spontaneously revealing the languages that were not completely foreign to him. This is something that might inspire authors of neuro-science-fiction . . . It is all the same more high-tech and more humane than the Pentothal, or truth serum, found in old detective films.

THE INTIMATE BRAIN

fMRI is beginning to affect all of society, and studies on the "social" side of the brain are beginning to flourish, some looking into the neuronal bases for dominance, or the molecular mechanisms for monogamy . . . The ten largest departments of psychology in the United States are equipped with fMRI scanners, and more than 200 scientific journals today publish articles related to this technology. At first these studies involved social interactions linked to reproductive behavior and parental care, and relatively simple cerebral mechanisms were purported to have been found. Other studies have involved, and are still looking at, abnormal social interactions, autism or schizophrenia. Another area of study involves the effects of social isolation and separation, which can be as devastating to our health as those of smoking or obesity.

fMRI reveals that a network of cerebral structures is involved in these normal or abnormal social behaviors. The "amygdala" at the base of our brain, for example, is involved in the recognition of social emotions (guilt, arrogance) or the sensation of fear (a student came to see me once at the end of class, livid: they had taken out his *amygdales* [also the word in French for "tonsils"] when he was a child. Rest assured, those *amygdales* are at the back of the throat, not in the brain!). The amygdala is closely connected to the ventromedial prefrontal cortex, which controls social judgment and

73

positive subjective interactions. As for love and feeling the loss of another, they are connected to the "striatum" and the "insula." Romantic attachment is associated with the "anterior cingulate cortex," whereas social exclusion is within the "ventral prefrontal cortex" of the right hemisphere. Cultural aspects, here more than elsewhere, intervene and must be taken into consideration. Thus one of my colleagues sought to identify in a population of young French adults the regions involved in pleasure and disgust. They were shown a series of random images ranging from a very pleasant face to horrible images of mutilated children or people with tumors on their faces. These images came from a database shared with an American research center. The hedonic nature, pleasant or unpleasant, was clear for the images with extreme characteristics. But the photo of a hamburger from a famous chain of fast-food restaurants, for example, was classified among the subjects of his study as negative, whereas it was assumed to have a positive connotation, at least for American subjects.[2]

Western music is based on scales constructed with very well-defined intervals (tones and half-tones), different scales having very precise relationships, "distances" between them, which Johann Sebastian Bach magisterially systematized in his *Well Tempered Clavier*. Passing from one tonality to the other, "modulation," creates all the richness of our music, and composers make use of it to express themselves, but within strict rules that constitute the "harmony." And it seems that these rules indeed exist within the depths of our brain . . . in particular in the "medial prefrontal cortex." If the rules aren't followed, even nonmusicians will sense they are hearing the wrong notes, which makes this region "scream" on an fMRI image (reference 4.17). As for the great shiver of delight felt by music-lovers listening to works like the intermezzo adagio of Rachmaninov's *Piano Concerto No.3*, or Barber's *Adagio for Strings*, it comes from the activation of centers of reward, motivation, and finally "pleasure," the prefrontal and orbitofrontal cortex, mesencephalic region, and so on, zones that are also activated by chocolate, sex, and some drugs, such as cocaine (reference 4.18).

And so with fMRI we are in the process of reviewing human behaviors and establishing a catalog of cerebral regions and their involvement in a number of these behaviors (most are not at all exclusive to humans) and in patho-

---

[2] It is interesting to note, by the way, that the taste of the hamburgers sold by this chain of fast-food restaurants throughout the world is very different from one country to another . . .

logical behaviors. Since these regions are often involved in multiple cognitive processes, they constitute networks capable of connecting in response to a given situation. These fMRI studies are gradually beginning to enter the field of psychiatry. Thus the region of the "fusiform gyrus" is not activated in an autistic person who looks at faces (autistic children have difficulty establishing visual contact with others), but this can simply correspond to a lack of interest or expertise in the recognition of faces, and not necessarily to a functional abnormality. Moreover, these patients in general have exacerbated expertise in other realms, such as calculation (some are "mathematical prodigies," and sometimes authentic mathematical geniuses) or memorizing impressive lists of items.

Imaging has modified our perception of mental illnesses, challenging psychological and neurochemical theories of the past. All the same, more so than in functional abnormalities of particular regions, the scientific community is currently interested in abnormalities of the neuronal circuits and the connections between these networks, which can be revealed through MRI, as we will see in the next chapter, and are associated with developmental and genetic factors. This new path of inquiry has given rise to hitherto unknown therapeutic approaches, such as the use of deep cerebral stimulation using electrodes implanted in the brain to modulate the activity of neuronal circuits. Very encouraging results with prolonged remissions have already been obtained in depressed patients for whom psychotherapy and medication are ineffective. The white matter of the "subcallosal cingulate region" is stimulated via electrodes placed (under MRI control) in the two hemispheres. Imaging studies also highlight the importance of reviewing classifications of psychiatric illnesses, such as depression, in fact, which covers many pathologies. It's a bit like a "fever" was in the days before the dawn of modern medicine (they even called them "fevers"); today a fever is only a simple symptom that quickly disappears with medication.

fMRI thus reaches into the deepest part of us, revealing who we are, from our genes, our history, and our environment. Education profoundly affects the functioning of our brain, as Stanislas Dehaene in our laboratory has shown: fMRI reveals that learning to read causes specific activity to occur in certain cerebral regions, as in the visual recognition of words, sometimes at the expense of other regions that become displaced, such as those associated with the recognition of faces (reference 4.19). Our socioeconomic status also has a profound impact on the functioning and the structure of our brain. Differences, modest but significant, observed in the linguistic abilities of school-age children distinguished by socioeconomic status, are correlated with clear differences in the lateralization of the inferior frontal gyrus of Broca's area. Children at the highest socioeconomic level are more lateralized on the left (reference 4.20).

We are also beginning to be able to prove the influence of our genes on the functioning of our brain thanks to MRI images. Our colleague Philippe Pinel has shown that the activation of regions that process language, in a population of normal subjects, can be modulated by the polymorphism of a single nucleotide (a basic element of the genetic code) within that population, in genes known to be associated with dyslexia (DYX2) or orofacial dyspraxia (FOX2) (reference 4.21). These results lead us to raise fundamental questions about what we truly are, a mixture of the expression of genes and exposure to the environment, and to wonder if, in the end, we are truly masters of ourselves and our decisions. Here, too, fMRI reveals many surprises . . .

DOES FREE WILL EXIST?

How does our brain decide between a rational, cognitive action and a moral, socioemotional action when such actions are in conflict with each other? Imagine you are hiding with your family in the basement of a house while the town is besieged by enemies ready to kill you. You hear the enemies arrive, but at the same moment your baby starts to cry and wails loudly. You cover its face with your hand to prevent it from being heard, but then it can't breathe. Should you take away your hand, allowing for the possibility that the enemy will hear your baby's cries and come to kill your entire family? Or should you keep it there and smother your baby to protect the other members of your family? The debate between a "utilitarian" attitude and a "moral" attitude isn't new, and is reflected in the dichotomy found in the theories of the philosopher John Stuart Mill, who promotes the "better position," and that of Immanuel Kant and his "deontologists," who maintain that certain moral limits must never be crossed, regardless of the circumstances. MRI shows that there is greater activation of regions such as the "dorsolateral prefrontal cortex" and the "anterior cingulate cortex," regions associated with abstract reasoning and cognitive control, than of cerebral regions involved in emotion (and the breach of morality), which enables us to predict a rational and utilitarian decision (reference 4.22). The philosophical controversy thus seems to have emerged from the competition between two sub-systems in our brain.

The economic consequences of these results have not gone unnoticed. What motivates the buying frenzy at the time of big sales? Many would like to be able to predict the decisions of a trader, or quite simply a buyer in a supermarket or on the Internet. With the use of credit cards and deferred payments, the buyer is in a certain way "anesthetized" from the pain of paying, and the pleasure of buying easily wins out. "Neuroeconomy" and "neuromarketing" (two very different concepts, in fact) have made their appearance, and no fewer than fifteen companies are requesting the use of fMRI to "test" their products and sales strategies. Showing men sports cars compared to less desirable cars activates the "mesolimbic system." Showing a favorite brand of a drink (beer for men, coffee for women) activates the "medial prefrontal cortex," but this does not necessarily indicate a preference for a product, since the price is not indicated. When subjects are shown an item for 4 seconds, then its price, again for 4 seconds, and they are finally asked to decide whether or not they would buy it 4 seconds later, the decision to buy can be predicted better from fMRI images than from criteria reported by the subjects! Thus, the activation of the "nucleus accumbens" (sensitive to a preference for a product) while the item is being shown, and the inhibition of the "insula" (sensitive to an excessive price) and the activation of the "medial prefrontal cortex" while the price is shown, predict the purchase of an item by a subject, either man or woman (reference 4.23). Soon perhaps, when we buy something on the Internet, we will be connected to our computer by a helmet lined with electrodes that will have replaced the mouse and the keyboard . . .

But who ultimately decides, "us" or our brain? You no doubt know that ambiguous image in which one can see either a vase or two faces looking at each other (figure 4.10). It is impossible to see the two images at the same time, because we necessarily see one or the other. If these images are shown to volunteers in an MRI scanner for an instant (a tenth of a second), they will see either the faces or the vase in an apparently unpredictable way. When a volunteer has seen the faces, fMRI shows that the specific region for recognizing faces is activated, which, after all, is normal. But what our team member Andreas Kleinschmidt discovered was more surprising (reference 4.24): the activation of the region for face recognition begins a few seconds *before* the image is presented to the subject. Clearly, he saw faces *because* that region was spontaneously preactivated! If that region is not in a state of pre-activation, the vase is seen (figure 4.10).

Our cerebral regions are in fact in a permanent functional state of fluctuation, without our being aware of it, beyond the realm of consciousness. The awareness of something at a given moment thus depends on the state of our brain at that moment. Is the impression of "freedom" in our judgments thus an illusion, like that ambiguous image? To find out more, another team asked volunteers, who were shown letters on a screen twice per second, to press a choice of buttons, either on the right or on the left, whenever they wanted to, and just when they wanted to. They simply had to remember the letter seen at the moment of their choice. To verify, a screen showing the last letters presented then appeared and the subjects had to recognize the letter retained by again pressing a button. An analysis of the resulting images reveals that the activation of a particular region of the "frontopolar cortex" *10 seconds* before the conscious motor response of the subjects, enables us to predict whether they are going to respond with the button on the right or on the left (figure 4.11; reference 4.25). fMRI even shows that the initially unconscious activation, the precursor to the motor response decision, is then relayed to the "precuneus," behind the parietal lobe where the motor response is constructed, before progressing to the supplementary motor area that finely tunes the motor movement, still unconsciously, a few seconds before the subject "is aware," that he *wants* to push the left button . . . So there are constantly many things happening in our brain, without it telling us about them.

At the Doors of Awareness

Now let's show our subjects a word, for example "lion," flashing on a screen for 30/1000 of a second. If nothing appears on the screen after the word, it is easily recognized and identified by the subjects. The fMRI images then show a complete network that is activated, visual and temporal areas linked to language, even the Broca area (the word is "on the tip of the tongue . . . "), and of course the region of the visual recognition of words. Now let's change the method of presentation a bit. This time the word, still shown for 30/1000 of a second, is preceded and followed by a screen showing other random images (what we call a "mask"). In this case, the word becomes "subliminal," and the subjects have no awareness of it. As proof, if we then ask them to choose a word among four that they "might have" seen, the subliminal word is chosen at random, with one chance in four of being picked. But what has their brain seen? With fMRI, from which (almost) nothing escapes, when we compare images of the preceding test to images obtained when no subliminal word was presented, we find visual activation (the brain has seen something . . .), activation in the region of the visual recognition of words (it has recognized that it was a word . . .) and even in Broca's area (the word almost went to the tip of the tongue . . .) (figure 4.12). And yet the subject is not at all aware of this!

In fact, the activation is too weak to have unleashed a conscious response, but the response is visible, and the word has even been memorized: if it is shown again, the responses will be even weaker, compared to the (subliminal) showing of a new word. These experiments, carried out by Stanislas Dehaene, suggest that the brain is constantly active, reacting to multiple input processed locally via a great number of elements and in cerebral subregions (reference 4.26). If "something important" happens, a central coordinator (which might be localized in the frontal lobe) connects the elements and amplifies their response, which makes that "thing" emerge in the realm of our consciousness. If we are busy at a task, for example, playing a video game, while others are busy around us, we ignore them and pay no attention to them. If one of them shouts because he has pinched his finger, we immediately turn our head: the spell has been broken (for us, as well as the other, by the way).

All these attempts to understand the mechanisms of our consciousness quite naturally lead to our wondering: to what does the apparent *absence* of consciousness, when we are asleep, are anesthetized, or in a coma, correspond? What is happening in our brain then? An English team asked this question in treating a 23-year-old woman who for 5 months had been in a confirmed vegetative state, exhibiting no response, following a severe head injury from a car accident.

"What is your name, Miss?" No response. And no response when she was pinched sharply . . . until the patient was placed in an MRI scanner and they saw the language regions light up . . .

"There is sugar and cream in the coffee," she was told. Immediately, there was activation in the temporal lobes, as in a normal, completely conscious subject (reference 4.27)! By contrast, they observed no response to nonverbal sounds. And if the phrase was coherent, but contained bizarre words, the Broca's area was more highly activated, as in a normal subject.

"Can you imagine, Miss, that you are playing tennis?" Immediately the supplemental motor region was activated.

"You are walking in your house . . . " Then the "parahippocampal gyrus," the "posterior parietal cortex," and the "lateral premotor cortex" were activated in a way that was indistinguishable from responses obtained among normal subjects who were asked to imagine the same thing. This woman, nevertheless in a complete vegetative state, thus understood what was being said to her and was capable of responding mentally, as no movement of her mouth or her body was possible (figure 4.13).

From all evidence, this young woman was aware of herself and her environment and cooperated with the team that was taking care of her. Since then the study has been repeated with a great number of subjects in vegetative states. Around 20 percent of them were capable of communicating in this way, via fMRI. This doesn't mean that the others did not have residual consciousness. On the one hand, fMRI is not always sensitive enough, as its images are sometimes unreadable due to uncontrolled movements; on the other, these patients might have minimal states of consciousness, fluctuating over time, and it was necessary to be there at the right moment to record them. But we can indeed see that this work opens up considerable therapeutic perspectives.

We can go further, a bit like with the patient in the book (and film) *The Diving Bell and the Butterfly*, who, stricken by a "locked-in" syndrome, is capable of choosing from among letters that are rapidly said to him one by one the ones he wants by moving his eyelid, until he manages to write a book: his story. For vegetative patients there is no longer any movement available with which to communicate; only thought revealed through fMRI indicates their mental existence. But it has also been possible recently to "dialogue" with some of them in a similar, much more complete way, with the complex fMRI signals being decoded by a computer. Who knows, perhaps Ravel might have one day "written," had the music of *Jeanne d'Arc* emerge from his head in this way! The same team has lately been able to obtain similar results from electrical signals (EEG) captured on the surface of the patient's head by having the machine placed beside the patient's bed, which makes the method much less expensive and infinitely simpler to use. However, EEG has been used up to now on thousands of coma patients, without anyone ever having found the slightest signal that might suggest that these patients were perhaps aware and were waiting to be able to communicate. It took fMRI to open our eyes and show us what we should be looking at.

Similarly, MRI was already several years old before it was understood that magnetized water molecules, through their song, could tell us the intimate story of our brain, opening a path to the very foundations of our consciousness. But those molecules still hide many other secrets, and it is through their dance this time that we are going to seek them out.

FIVE

THE BRAIN PROBED THROUGH WATER MOLECULES

Henry didn't feel well that evening. At the restaurant where he was celebrating his seventieth birthday with friends, he was having a lot of trouble using his right hand; it seemed heavy, as if it were asleep. Yet he didn't feel any pain. His friends didn't notice anything, except when he started having trouble talking—he couldn't find his words. At the end of the meal, he was talking bizarrely and had trouble walking: his right leg was dragging. Maybe he had drunk too much? He rarely did that. They took him home; he would be better tomorrow!

The next morning, Henry couldn't get out of bed, he couldn't even turn over. His right leg and arm no longer obeyed him. His speech was completely incoherent. The doctor came, but it was too late: like 800,000 people every year in the United States, Henry was the victim of a cerebral vascular accident (CVA), or stroke. A blood clot from his heart or an atheroma on the wall of the artery in his neck had detached and traveled into one of the arteries of his brain, and had become blocked. The cerebral tissue that depends on blood was no longer irrigated, and when deprived of oxygen and sugar, it ceased to function. In general, symptoms can vary greatly depending on the region of the brain affected. Henry escaped death, which strikes 20 percent of patients who suffer a stroke (130,000 people per year in the United States), the third cause of death after cancer and cardiovascular disease, but he remained handicapped, like 75 percent of patients who are victims of a CVA, paralyzed on the right side and aphasic, able only with great difficulty to communicate with those around him. Granted, with rehabilitation some progress is possible, but he would spend the rest of his life between a wheelchair and his bed. Like one patient in three, he would no longer be independent and would need help for all daily tasks.

However, Henry could have woken up as if nothing had happened. Let's go back to the evening before, at the restaurant. Seeing his bizarre behavior, his friends instead realize that he is having a stroke. They call 911 and an ambulance takes him immediately to the emergency room, where he sees a neurovascular specialist. In the hospital he immediately undergoes an MRI, in particular, a "diffusion MRI." The diagnosis is clear: a white spot shows the cerebral region involved and specifies its localization in the brain (figure 5.1), in the left hemisphere, which explains his difficulty with language and the paralysis on the right side. Then begins a race against the clock: every minute that passes means another 2 million neurons are lost ("time is brain"). It's been less than 6 hours since the symptoms began; this is good, doctors can still try thrombolysis, injecting him with medication to dissolve the clot. Can they call it a miracle? In almost one patient out of two, in any event, their symptoms disappear immediately, and completely, within a minute. No more paralysis, no more aphasia. Rewind the tape . . . as if nothing had ever happened.

That scenario is not at all in the realm of science fiction. However, unfortunately, in 2014 only a very small fraction of individuals stricken with a CVA reach the hospital, much less a neurovascular emergency department in time. Yet we could do much better. While about one patient out of 100 could benefit from thrombolysis, it could be effective in 15 times more of them. Thus there is much work to be done to make the symptoms known to the public, to better arrange the transfer of patients with a suspected CVA to a hospital, and, just as it used to be the case for a myocardial infarction, to develop neurovascular emergency centers having MRI scanners accessible in the event of such an emergency. But even today the outcomes of some patients are very positive in part thanks to what MRI, and especially diffusion MRI, has brought to health care. For CVA such a possibility wasn't even conceivable only 20 years ago.

How has diffusion MRI become an incontrovertible tool in medical practice? MRI arrived in France at the beginning of the 1980s, when I was a medical resident and a PhD student in physics. The first images produced were of incredible "beauty." However, although anomalies in the brains of patients became much more visible than via the x-ray scanner, we could often not put a "label" on them, nor identify them, which was truly frustrating. This is sometimes still true today. This is because MRI images have a spatial resolution of around 1 millimeter, whereas the theater of the pathological and physiological processes is situated on the cellular level, on a scale that is around a thousand times smaller, that of the micron. This is why doctors, in order to make a precise diagnosis after having detected a lesion on the images, must then perform a biopsy—that is, remove the suspicious tissue and study it under the microscope. Although this is common for the diagnosis of cancer in some organs, such as the breast, it is obviously much more problematic when dealing with the brain: the skull must be opened and . . . the surgeon must aim the biopsy needle very well, as the brain is a particularly fragile organ. So brain biopsy is rarely performed, unless there are very specific reasons. The question I asked myself at the time was whether it would be possible to obtain information on a microscopic scale from images with a millimeter of resolution.

EINSTEIN'S VISIONS

Einstein raised that sort of question in 1905 within an entirely different context. The early 1900s was an extremely fertile period, in art as well as the sciences, and 1905 is considered to be the *annus mirabilis* (year of miracles) for Einstein. While he was completing his thesis in physics, he published an article on the photoelectrical effect that served as a foundation for quantum mechanics and earned him the Nobel Prize sixteen years later. The article on the theory of relativity and the one on the equivalence of mass and energy ($E = mc^2$), published the same year, were not immediately understood or accepted by the scientific community, nor was an article on molecular diffusion, which was the subject of his thesis (reference 5.1). At that time scientists knew of the existence of "Brownian motion." Robert Brown (1773–1858), a Scottish botanist, had observed under the microscope that grains of pollen were permanently agitated in every direction. At first he thought it was the secret of life ("it's my little secret," he confided in the young Charles Darwin, who had come to visit him right before he left on his world tour), but he quickly saw that this motion, which was also seen in inorganic particles, was a ubiquitous physical phenomenon in the microscopic world.

On the other hand, the phenomenon of "diffusion" was also known, in particular through the work of the Dutch chemist Jacobus Henricus van 't

Hoff (Nobel Prize in Chemistry in 1901), a phenomenon that enables two liquids of different nature or concentration to mix together (a phenomenon also called "osmosis") more or less rapidly. Thus sugar, once dissolved, can blend with our coffee (it takes some time, as diffusion is a relatively slow process, and so we generally cheat by using a spoon). The diffusion phenomenon is clearly visible in our macroscopic world.

Einstein's idea was to link the two phenomena, Brownian motion and diffusion. His intuition was that the (macroscopic) process of diffusion should be explained by the (microscopic) Brownian motion of atoms (or molecules), which, through random movement (like that of a drunk who tries to go home, stumbling), at the end of a certain time, ultimately fill the entire space in which they are contained and blend with each other. This Brownian motion is the result of the repeated knocking together of atoms or molecules in perpetual movement. However, at that time, the existence of independent atoms or molecules was far from being established. Atomic theory in particular, which explained heat through the kinetic agitation of atoms, was the subject of fierce battles among scientists. Legend has it that Ludwig Boltzmann, the Austrian physicist and founder of statistical mechanics and statistical thermodynamics, partisan of atomic theory, killed himself following the attack from his colleagues.

Einstein's objective was to prove the existence of atoms by demonstrating, using a physical model—equations—that the (visible) phenomenon of diffusion could be explained perfectly by the (invisible) movement of atoms. Thus in his physics thesis and his article devoted to diffusion, Einstein established that the random path effected by atoms (or molecules) in a given amount of time is directly linked to the diffusion coefficient (a "macroscopic" parameter that characterizes the rapidity of the process of diffusion for a given molecule) observed macroscopically. Furthermore, he showed how this diffusion coefficient is determined microscopically, among other things, by the size of the atom or the molecule, as well as by temperature (reference 5.1). This theoretical correspondence between the microscopic and the macroscopic worlds was verified experimentally by Jean Perrin, who once and for all established the existence of atoms, which earned him the Nobel Prize in Physics in 1926. Jean Perrin, using Einstein's equation and the diffusion coefficient, was also the first to determine the size of the water molecule: around 3/10 of a millionth of a millimeter.

Other scientists, such as Raymond Poincaré and Louis Bachelier, had been interested in Brownian motion and random movement before Einstein, but on a mathematical scale. Bachelier's 1900 thesis on the theory of speculation became a great classic, explaining fluctuations in financial markets and those in human behavior that followed similar laws! It is striking to realize that Einstein, throughout his work, was able to access realms on a scale that is completely inaccessible to our senses and uncover the laws that govern them, laws that also apply to the world on our own scale, whether we are looking at space through Brownian motion and diffusion, photoelectric effects (establishing the quantum nature of those effects), or time, reasoning so to speak at the speed of light to establish the physical laws (relativity) that govern our everyday lives.[1]

NMR Sensitive to Diffusion

But let's return to MRI. The idea that came to me in the middle of the 1980s was that, if it were possible to determine the diffusion coefficient of water on the millimetric scale of MRI images, it might perhaps be possible from that to deduce what happens on the microscopic level. A stroke of luck: the diffusion coefficient of water at the temperature of the brain (37°C) is such that, following Einstein's equation, water molecules diffuse over an average distance of around 10 microns for the duration of the acquisition of MRI signals (a fiftieth of thousands of seconds), or on exactly the scale sought! Water molecules could in a certain sense be seen as spies, their Brownian motion of random movement being slowed by the presence of obstacles such as cell membranes, fibers, or macromolecules of biological tissue (figure 5.2).

Thus the diffusion coefficient of water in a tumor made of cells that are different from the tissue in which it is found should differ from that in the neighboring tissue. But to show this we first had to encode the microscopic movement of diffusion in the MRI signal of the images—that is, create a new contrast in the image where, at each point, the diffusion coefficient of water appears, which was no small task. Granted, I was not starting from scratch. Indeed, shortly after the invention of NMR in 1946, it was established that the signals were "naturally" parasitized by diffusion. The reason for this was that the magnetic field of the first NMR machines was not very homoge-

---

[1] The precision of the GPS system that enables us to find our way depends upon it, for example.

neous. The atoms or molecules in motion (Brownian) thus saw the magnetic field change constantly, which was translated by small fluctuations in their frequency of resonance, which distorted the signals.

We then tried to render the signals *less* sensitive to diffusion, which was seen to be a pollution—for example, by acquiring them in a series of staggered, small intervals of time that were short enough so that the molecules didn't have much time to move. In the 1960s the American physicists E. O. Stejskal and J. E. Tanner showed that it is possible to measure precisely the diffusion of molecules by imposing strong (controlled) variations in the magnetic field in time and space (reference 5.2). For a brief moment the magnetic field is rendered highly variable following a given direction (we call this the impulse field "gradient"): the field passes rapidly, for example, from a high value on the left to a low value on the right. Water molecules thus "see" a different magnetic field depending on their position. A bit later (some dozens of thousands of a second), the magnetic field is again modulated for a short instant, but in the opposite direction. If the molecules haven't moved, the effect is exactly cancelled out, the excess of the field perceived the first time being exactly compensated for by a deficit in the field the second time. Conversely, if the molecules have moved, due to diffusion and their random movement, they are found in different positions and the effect of the variation of the perceived field is no longer exactly compensated for: those molecules in some way conserve a "memory" of their displacements (figure 5.3).

Since the movement of diffusion is by definition random and there are a considerable number of molecules, the result is a beautiful cacophony (an "interference") that partially destroys the NMR signal. This amputation of the signal is precisely determined by the path of the molecules following Brownian motion and thus by the diffusion coefficient following Einstein's equation, as well as by the temporal and spatial characteristics of the variations of the magnetic field being used. Since these characteristics are known, we can deduce the diffusion coefficient.

From NMR to MRI: Diffusion . . . Confusion

My contribution to the field has been to establish the method that enables a combination of the NMR principle and the processes for acquiring MRI signals in order to obtain images in which, at each point, the signal allows us to determine the diffusion coefficient of water. That molecule is at the origin of the MRI signal and, in addition, is of great importance to us (references 5.3, 5.4). This had never been achieved before. My first images were produced in 1984 using an MRI scanner of 0.5 tesla from the French company Thomson-CGR.[2] A few technical difficulties had to be ironed out, the main one involving the (involuntary) movements of the patients' heads. The images obtained of my own head, then those of my colleagues, and finally of patients, were very encouraging: new contrasts appeared! Thus in 1985 I had the honor of showing the very first images of water diffusion in the human brain at the meeting of the international society of MRI in London, and then at the annual meeting of the North-American Radiological Society (RSNA) in Chicago.

But the reception of the scientific community was rather cool: some people remained skeptical, denying the possibility that MRI images could really be sensitive to microscopic molecular displacements. Others found the method too complex (I had also shown how diffusion MRI could be used to obtain images of "perfusion," movements of water circulating in small blood vessels [reference 5.5]). Wanting to tease me some colleagues started wearing t-shirts with "diffusion, perfusion . . . " printed on the front, and ". . . confusion" on the back! For several years I continued to publish on my method and its applications, and my stubborn Breton insistence ultimately paid off. Today, diffusion MRI has become an incontestable method in radiology and is featured on almost all MRI scanners throughout the world. *Google Scholar* cites 613,000 references on the subject at the beginning of 2014.

What really launched diffusion MRI onto the medical scene was a surprising discovery made in 1989 by one of my colleagues, Michael Moseley, then at the University of San Francisco. Mike was trying to understand the mechanisms of cerebral ischemia, when a region of the brain is no longer irrigated due to a clot blocking a small artery, the cause of stroke. To do this he used an animal, a cat, in which he tied off a cerebral artery (this method was classic at the time), and he then studied the effects by acquiring MRI images of the animal's brain. Since I had shown that diffusion MRI was also sensitive to perfusion, Mike quite naturally used it. But to his great surprise, he observed that it was the diffusion of water that was altered in the ischemic

[2]CGR (Compagnie Générale de Radiologie) was bought by General Electric in 1987, which enabled GE to later become the first manufacturer to offer diffusion MRI on its MRI scanners.

cerebral region (reference 5.6): the Brownian motion of water was slowed by 30 to 50 percent in the dying region deprived of blood! The effect was clearly visible at the very onset of the ischemia.

The reason for this slowing of water diffusion remained obscure, but the news very quickly got around: diffusion MRI enabled us to "see" the ischemia at a very early stage, when the effects were still reversible if the blood flow was restored, which no other imagery technique enabled up to then. X-ray scanner images, for example, became abnormal only after many hours, when the tissue was already destroyed. The potential was thus enormous but remained to be demonstrated in humans, which wasn't easy. First, patients who were victims of a cerebral vascular accident (CVA) were not sent to the emergency room at that time, because there was no exam enabling the diagnosis to be confirmed, nor was there special treatment available (thrombolysis had just been approved only for myocardial infarction). Only a few research centers had a handful of patients. And to validate the potential of diffusion MRI there had to be an MRI scanner available in the emergency room, one equipped with the technology that I had just invented. Furthermore, the method was still very experimental: it took almost 20 minutes to get the images, and those were not always of the required quality, especially when patients were agitated.

I was working in the United States at the time, at the National Institutes of Health (NIH). Robert Turner and I were using the method of ultrarapid imaging through "echo-planar" for functional MRI. Very quickly it seemed obvious to us that the echo-planar method might also be used to obtain images of diffusion, which we had demonstrated in an animal, then in a human with the close collaboration of the General Electric (GE) teams. Diffusion MRI through echo-planar enabled us to obtain images in less than a minute, images that were moreover insensitive to the movements of the heads of patients (reference 5.7). GE installed the method in a few select hospitals, such as Massachusetts General Hospital in Boston. In 1993 it was established that diffusion MRI could enable us to differentiate the cerebral regions—also in humans—afflicted by a vascular accident within the first hours of the onset of the event (figure 5.1; reference 5.8).

Thrombolysis was in the process of being evaluated, and diffusion MRI enabled physicians to objectify the conditions of its use, proving that its efficacy was maximal during the first three hours after the event. It also enabled them to follow the progress of the patient after the acute episode (with a return to normal diffusion in the afflicted area). The progress achieved through diffusion MRI was determinative, and thrombolysis was officially approved by the Food and Drug Administration (FDA) for stroke in 1996. Following these results, diffusion MRI was proposed by every MRI scanner manufacturer, and became an incontrovertible clinical tool at the end of the 1990s. And so today, a large, but still insufficient, number of patients, victims of stroke, can escape paralysis, aphasia, or being handicapped for the rest of their lives.

In addition, it has been shown that the slowing of water diffusion is associated with the swelling of dying cells, although the mechanism connecting the two remains unclear. The possibilities offered by diffusion MRI continue to grow: the method enables us to identify, within the first hours of an event, regions that have suffered but are not yet destroyed, and to predict which ones will be destroyed if nothing is done. In certain conditions a possible restoration through thrombolysis beyond the crucial first 3 to 6 hours lengthens the therapeutic "window" and gives these patients more chances for recovery.

DIFFUSION AND CANCER

Water diffusion MRI is also used outside the brain—in particular, in the diagnosis of cancer—and its use is increasingly competing with traditional approaches. The classic diagnostic method, in fact, relies on positron emission tomography (PET), which we discussed in chapter 3. Instead of injecting radioactive water to explore cerebral function, we administer to the patient in whom we suspect cancer or metastases another molecule, fluorodeoxyglucose (FDG) made radioactive through its atom of fluoride. Since the radioactive half-life of fluoride is 2 hours, we have a bit of time, and radioactive FDG can be prepared in specialized labs and delivered by shielded trucks to hospitals equipped with a PET camera, under the condition that there is no delay. Deoxyglucose is unique in that it is a "false" sugar because it lacks an atom of oxygen. But to cells it resembles the real thing almost exactly, and cells that lack energy grab on to it eagerly. Too bad for them! Indeed, unlike glucose, FDG cannot be broken down to produce energy, and it is accumulated in the cell, where it is toxic. Since FDG is radioactive, the PET camera

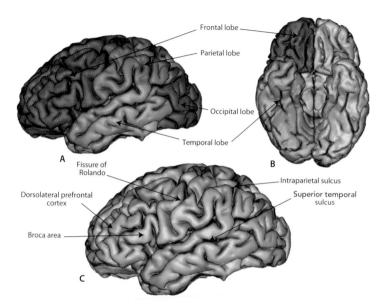

Figure 1.1. 3D rendering of the anatomy of the brain from MRI images. **A**: Lobes of the left hemisphere. **B**: View from below. **C**: Sulci and structures of interest.

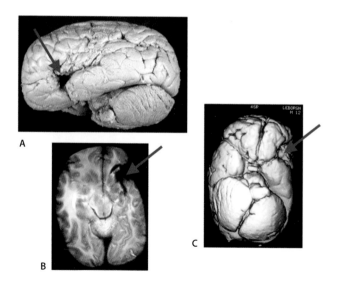

Figure 1.2. The brain of M. Leborgne (Broca's aphasic patient). **A**: Photo of the brain. **B**: Axial MRI section. **C**: 3D reconstruction from MRI images. The lesion is quite visible (arrow).

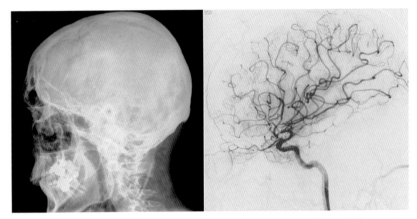

Figure 1.3. A: Cranial radiography. **B**: Angiography showing the opacification of the vessels inside the skull.

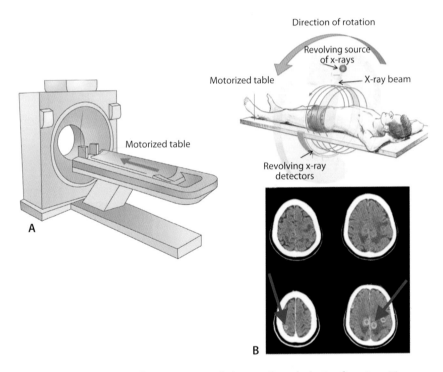

Figure 1.4. A: Functioning of an x-ray scanner. **B**: Sections from the brain of a patient. The contrast between white and gray matter is not highly visible. However, lesions (metastases, indicated by the arrows) appear clearly after the injection of iodine, which opacifies the blood vessels.

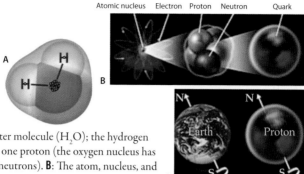

Atomic nucleus   Electron   Proton   Neutron   Quark

A

B

C

N   N

Earth   Proton

S   S

Figure 1.5A. A: The water molecule (H_2O); the hydrogen nucleus contains only one proton (the oxygen nucleus has eight, as well as eight neutrons). **B**: The atom, nucleus, and components. **C**: The proton carries a magnetic moment, just like the Earth, connected to its "rotation."

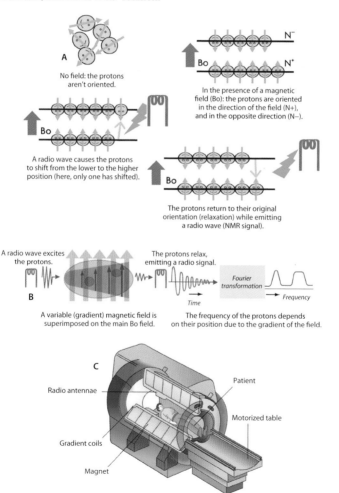

A

No field: the protons aren't oriented.

Bo   N⁻   N⁺

In the presence of a magnetic field (Bo): the protons are oriented in the direction of the field (N+), and in the opposite direction (N−).

Bo

A radio wave causes the protons to shift from the lower to the higher position (here, only one has shifted).

Bo

The protons return to their original orientation (relaxation) while emitting a radio wave (NMR signal).

A radio wave excites the protons.

B

The protons relax, emitting a radio signal.

Fourier transformation

Time   Frequency

A variable (gradient) magnetic field is superimposed on the main Bo field.

The frequency of the protons depends on their position due to the gradient of the field.

C

Radio antennae

Patient

Motorized table

Gradient coils

Magnet

Figure 1.5B. A: The principle of NMR. **B**: Principle behind MRI (localization of NMR signals through gradients in the magnetic field). **C**: Anatomy of an MRI scanner.

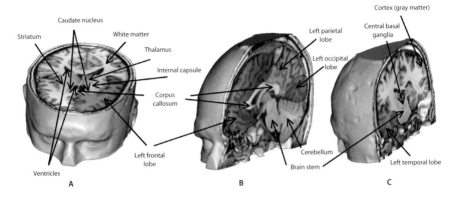

Figure 1.6. 3D rendering of the head and the brain from MRI images. The contrast between gray and white matter is clearly visible, with many details. **A**: Axial section. **B**: Sagittal section. **C**: Coronal section.

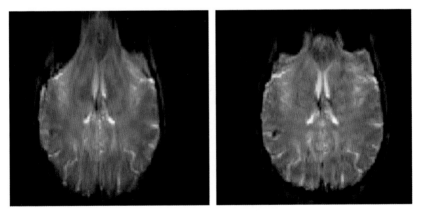

Figure 2.1. Axial MRI sections showing geometrical deformations at the front of the brain caused by the air present in the sinuses of the face.

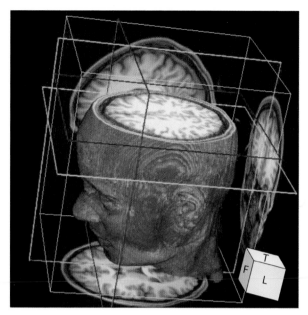

Figure 2.2. 3D reconstruction of the head and the brain from MRI images, and axial (below), coronal (behind), and sagittal (in the background) projections. **F** = front; **L** = left; **T** = top.

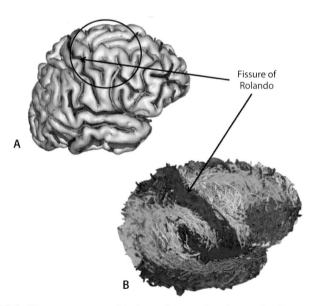

Fissure of
Rolando

A

B

Figure 2.3. A: 3D reconstruction of the brain showing the sulci in color. **B**: Superposition of the brains of multiple subjects, showing the great variability in the position and the shape of the sulci.

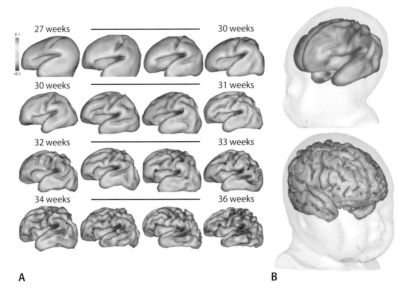

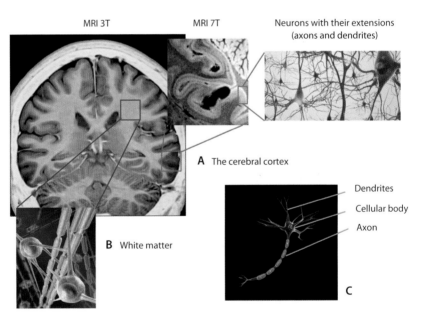

Figure 2.4. A: 3D reconstruction of the brains of premature babies showing (folding) development depending on the number of weeks of the pregnancy. **B**: Positioning in the head.

Figure 2.5. Axial MRI section of the brain at 3 teslas and a zoom view at 7 teslas showing **A**: the cerebral cortex (layer of gray matter covering the brain, and made up of six layers of neurons); **B**: white matter, made up of bundles of axons that ensure intracerebral connections; and **C**: the anatomy of a neuron.

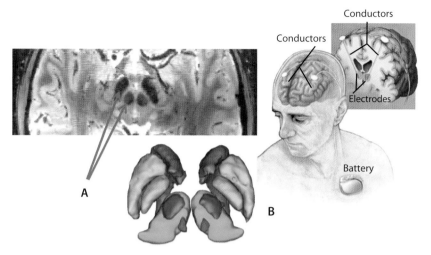

Figure 2.6. A: Basal ganglia visible on an axial section of a brain from 7 tesla MRI with their 3D reconstruction. **B**: Diagram of the principle behind deep stimulation of the basal ganglia by implanting electrodes.

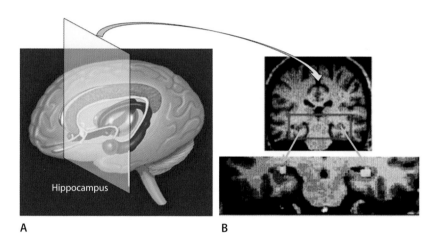

Figure 2.7. A: Localization and shape of the hippocampus (left and right). The hippocampus has a key role in memory and spatial recognition. **B**: MRI sections showing hypertrophy of the hippocampus in London taxi drivers.

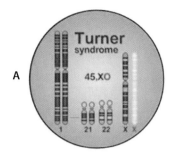

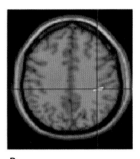

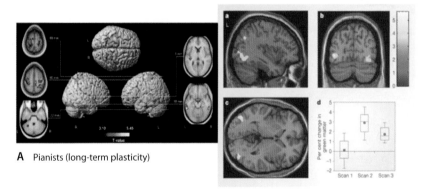

Figure 2.8. A: Turner syndrome is caused by the partial or complete deletion of an X-chromosome. **B**: Axial MRI section of the brain showing a loss of gray matter in the right parietal cortex. **C**: Reconstruction of sulci showing their defects in patients.

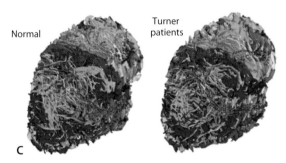

A Pianists (long-term plasticity)

B Jugglers (short-term plasticity)

Figure 2.9. A: MRI images showing in color regions (motor, auditory) that are more developed in pianists (long-term plasticity). **B**: Development of regions involved in seeing movement after training in juggling.

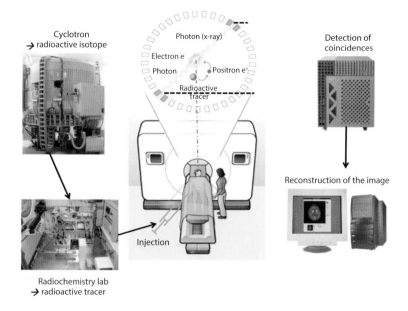

Figure 3.1. Principle behind the PET camera.

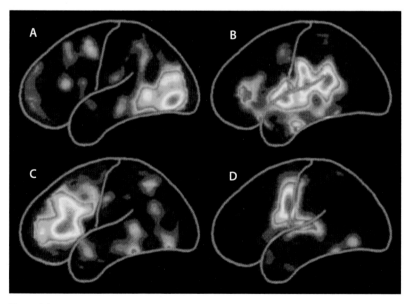

Figure 3.2. Historic image of the visualization of the activation of cerebral regions with PET. **A**: Visual. **B**: Auditory. **C**: Language/production of words. **D**: motor.

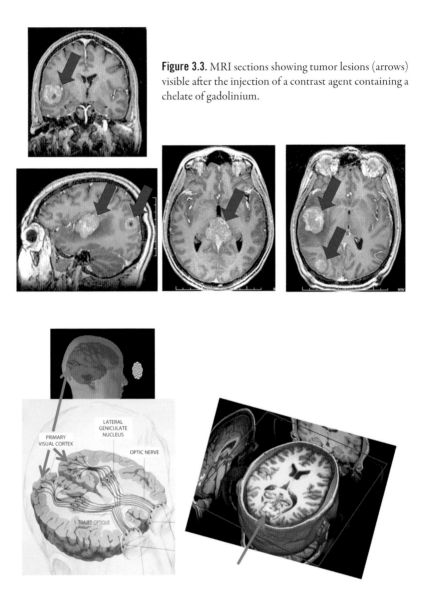

Figure 3.3. MRI sections showing tumor lesions (arrows) visible after the injection of a contrast agent containing a chelate of gadolinium.

LATERAL
GENICULATE
NUCLEUS

PRIMARY
VISUAL CORTEX

OPTIC NERVE

TRAJET OPTIQUE

Figure 3.4. Functional MRI (fMRI): The image of the checkerboard detected on the level of the retina is projected by the optic nerves onto the occipital cortex at the back of the brain. The degree of activation of the visual cortex, detected by fMRI, is represented by a scale of colors (arrow).

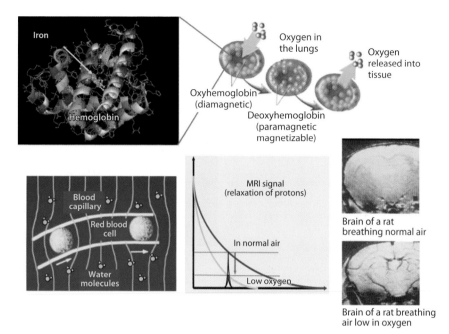

Figure 3.5. The Ogawa experiment: The hemoglobin of red blood cells contains an iron atom that is magnetizable. The circulation of deoxygenated red blood cells disturbs the magnetization of water molecules around them, leading to a lowering of the signal (the vessels become visible).

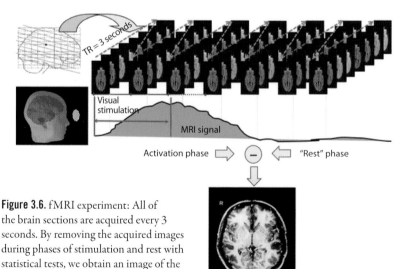

Figure 3.6. fMRI experiment: All of the brain sections are acquired every 3 seconds. By removing the acquired images during phases of stimulation and rest with statistical tests, we obtain an image of the activated regions that are seen by colors (here, visual activation).

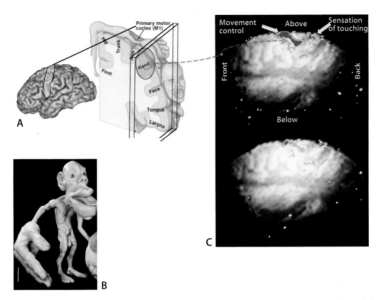

Figure 3.7. A: Cerebral representation of the regions of the body on the surface of the motor cortex (homunculus). **B**: The homunculus in 3D, illustrating the importance of the hand. **C**: fMRI image taken while the subject is tapping his fingers (top), or while imagining the movement (bottom).

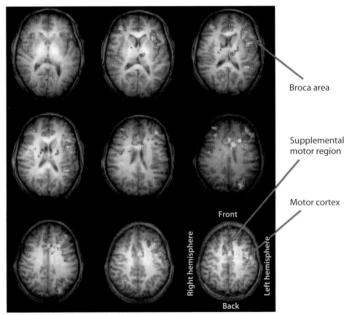

Figure 3.8. Mental production of words in a 12-year-old. Several regions are activated (including the Broca area), principally in the left hemisphere in this right-handed child, the brain being seen as if from above.

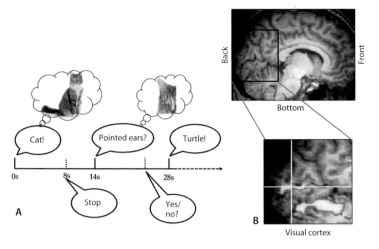

Figure 4.1. A: Protocol of mental imaging. The subjects must imagine an animal visually for 8 seconds and then respond mentally to a question about the animal. **B**: fMRI shows that the primary visual region (connected to the real seen world) is activated.

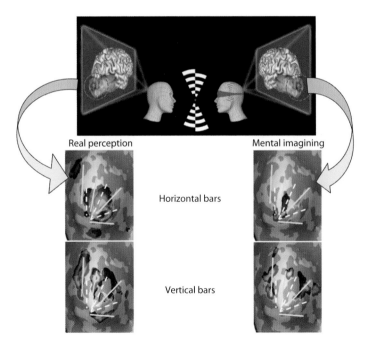

Figure 4.2. Left: Vertical and horizontal sectors seen by subjects give different motifs of activation on the visual cortex. Right: In the dark, subjects imagine the horizontal or vertical sectors. From fMRI images, it is possible to guess the orientation imagined by the subjects.

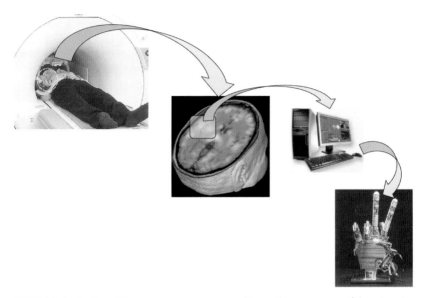

Figure 4.3. Activation of the motor cortex corresponding to the movements of the subject's hand in the MRI scanner is analyzed in fMRI images by a computer. The computer sends electrical signals to an artificial hand, which reproduces the movements of the subject at a distance.

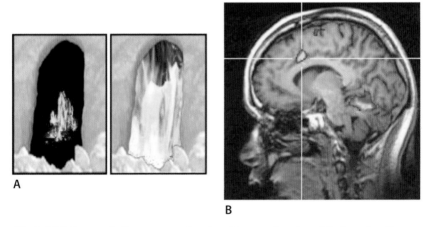

A

B

Figure 4.4. A: Images of a flame presented to the subject as a function of the intensity of activation detected through fMRI in the targeted region. **B**: Sagittal fMRI section showing activation in the targeted region (anterior cingulate rostral cortex).

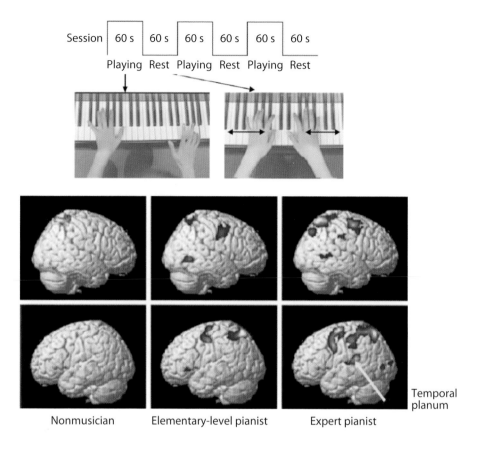

Temporal
planum

Nonmusician   Elementary-level pianist   Expert pianist

Figure 4.5. Network of regions activated by viewing the movement of the hands of a pianist playing a piece. In experienced players, this sight activates, among others, the premotor regions, the inferior frontal gyrus, the parietal lobule, and above all the temporal planum.

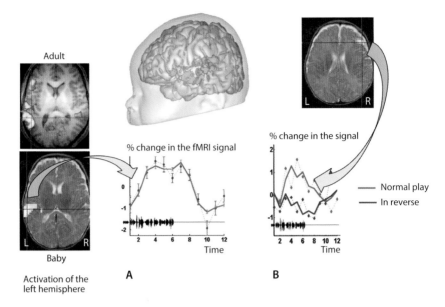

Adult

Baby

Activation of the
left hemisphere

% change in the fMRI signal

% change in the signal

Normal play
In reverse

Time

Time

A

B

Figure 4.6. A: Listening to a specific voice activates in a 2-month-old baby the left auditory regions, as in an adult. **B**: In the right frontal cortex, listening to the normal recording provokes activation; the recording played in reverse provokes no response.

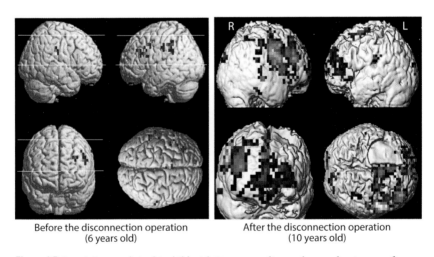

Before the disconnection operation
(6 years old)

After the disconnection operation
(10 years old)

Figure 4.7. Imagining words in this child with Rasmussen disease shows a dominance of language on the left—on the side of the lesion that will, however, be operated on. Four years later, the child recovered language and the right side is activated.

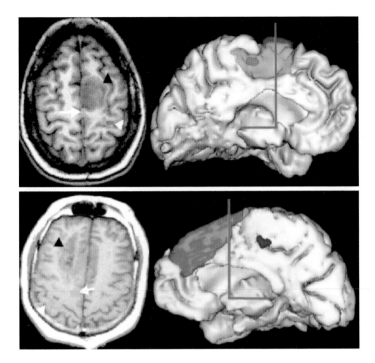

Figure 4.8. Top: The tumor (in green, black arrow) encompasses the supplemental motor area (SMA, in red, white arrow). The excision of the tumor with the SMA will lead to a motor deficit that will be restored a few weeks later. **Bottom**: The SMA not being in the tumor, no deficit is observed.

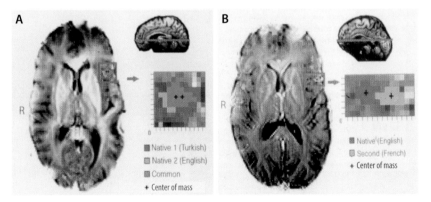

Figure 4.9. A: Activation of the Broca area by imagining words in Turkish or English is super-imposable in this subject raised speaking both languages. **B**: The Broca area here, however, is divided in two in an American who learned French as a second language in junior high school.

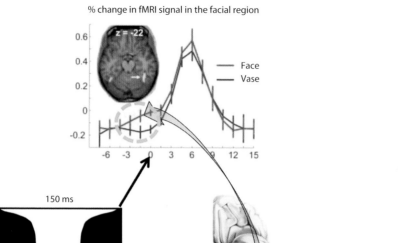

% change in fMRI signal in the facial region

z = -22

Face
Vase

150 ms

Facial area

Figure 4.10. Vase or faces? fMRI reveals that if the region specific to the area of facial recognition is activated (spontaneously) just before the image is shown, we will see the faces. If not, the vase is perceived.

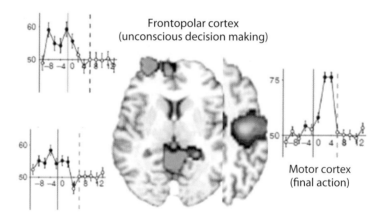

Frontopolar cortex
(unconscious decision making)

Motor cortex
(final action)

Precuneus and posterior cingulate cortex

Figure 4.11. Whereas the subject thinks he is freely choosing the moment when he pushes a button (motor cortex, orange area) to select a letter, fMRI reveals that the decision originates (unconsciously) close to 10 seconds earlier in the frontopolar cortex (in green). The red bar indicates the moment when the subject is aware of his decision.

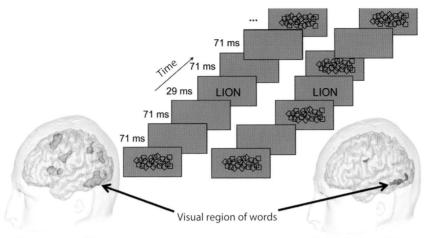

Figure 4.12. Displaying a word for a very short time is sufficient to activate a complete network of regions tied to language. If the word is preceded and followed by a mask made of random characters, it is no longer perceived (subliminal stimulus) but is detected all the same, among others, by the visual region specific to words.

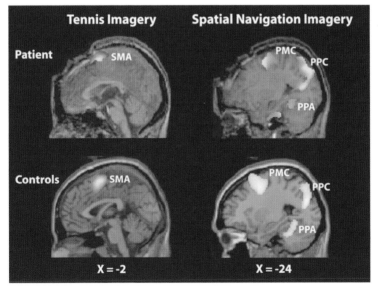

Figure 4.13. fMRI images obtained from a 23-year-old patient in a vegetative state. When the patient was asked to imagine playing tennis or walking in her house, cerebral activation was detected, very close to that obtained in conscious control subjects.

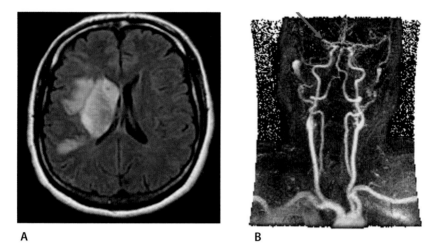

A   **B**

Figure 5.1. A: Diffusion MRI showing the region deprived of blood (ischemia) in the process of dying (white spot). **B**: MRI angiography showing the blood vessels and the occlusion of the right middle cerebral artery (red arrow) responsible for the lesion.

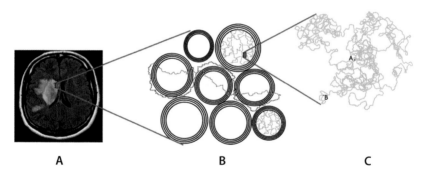

A   **B**   **C**

Figure 5.2. MRI showing a lesion (white spot) on a macroscopic scale of the image. **B**. Microscopic movement of water diffusion in the tissue, blocked in the cells (green) going through them (red), or moving around them (blue). **C**: Diffusion of water molecules on a molecular level, showing the random path between points **A** and **B**.

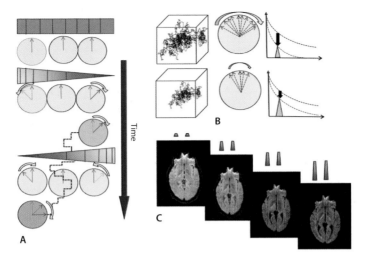

Figure 5.3. A: Diffusion encoding. Under the effect of a field gradient, the frequency of protons accelerates or slows, an effect that is upset by the following gradient unless the protons have changed place in the meantime due to diffusion. **B**: The random displacement of molecules produces a spreading of phases among protons, the source of a loss of signal that is all the greater when diffusion is rapid and the gradient intense. **C**: If we increase the intensity of the field gradient, the contrast of the image becomes increasingly sensitive to water diffusion.

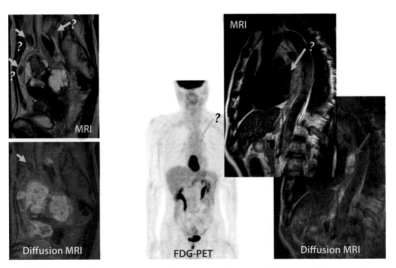

Figure 5.4. A: Serious adenocarcinoma and metastasis (cancer of the uterus). Images of suspect lesions (blue arrows). Diffusion MRI confirms the metastatic nature of two of them (red arrows), but not the third (normal intestinal loop, green arrow). **B**: Nodal metastasis (cancer of the esophagus). Diffusion MRI shows the metastasis that is not visible in MRI or PET

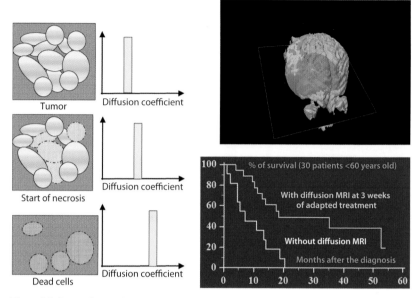

Figure 5.5. A: The death of cancer cells through chemotherapy causes a rise in diffusion. **B**: Increase in diffusion (green zones) due to an effective treatment. **C**: Improved survival curve for patients whose treatment has been modulated by diffusion MRI.

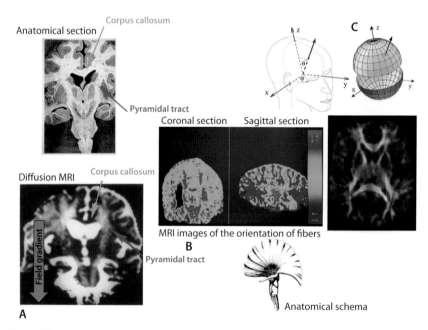

Figure 5.6. A: The corpus callosum, perpendicular to the gradient, appears in black, and the pyramidal tract, parallel to the gradient, in white. **B**: First-ever image of the orientation of fibers of white matter. **C**: 3D encoding in color.

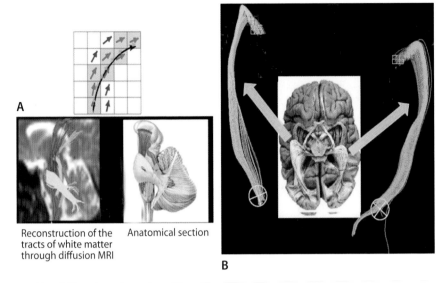

Reconstruction of the
tracts of white matter
through diffusion MRI

Anatomical section

A

B

Figure 5.7. A: Reconstitution of the trajectory of tracts of white matter in the brain stem obtained by connecting point-by-point the orientation of the fibers found through diffusion MRI. **B**: Reconstitution of optic radiations through diffusion MRI.

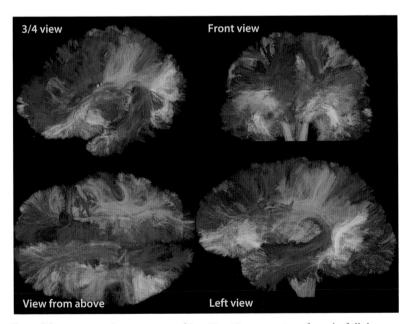

3/4 view   Front view

View from above   Left view

Figure 5.8. Automatized reconstitution (BrainVisa/Connectonist software) of all the tracts of intracerebral connections of a subject through diffusion MRI images. Each color corresponds to an identified tract.

Figure 5.9. Images of the principal tracts of fibers visible in 2- to 4-month-old babies. Most of the large tracts of adults are already present, although their functional maturity has not yet been reached (the myelination process of fibers).

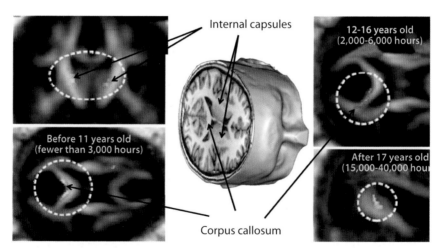

Figure 5.10. Correlation between the organization of fibers of white matter and the number of hours of piano practice at different ages in eight pianists. **A**: Internal capsules and isthmus of the corpus callosum. **B**: Splenium of the corpus callosum. **C**: Right arched tract.

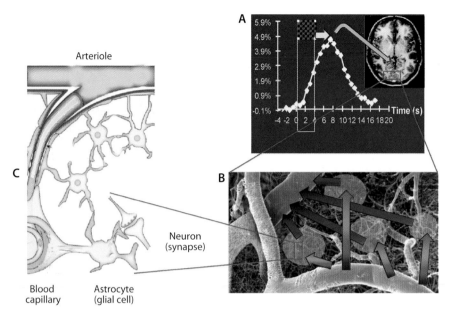

Figure 6.1. A: The peak of the BOLD vascular response is delayed by several seconds after the visual stimulus has stopped. **B**: In the cortex, different territories are irrigated by the same arterioles and drained by the same venules. **C**: The vascular response to neuronal activation is piloted by astrocytes.

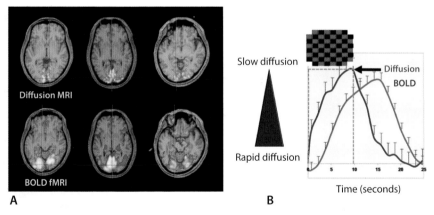

Figure 6.2. A: fMRI images of the visual cortex. Activation visible through diffusion appears more precisely (cortical ribbon) than that of BOLD. **B**: The response observed in diffusion (diffusional slowing) is in advance of the BOLD response (vascular), which indeed terminates at the moment the stimulation stops, whereas the BOLD response continues for several seconds.

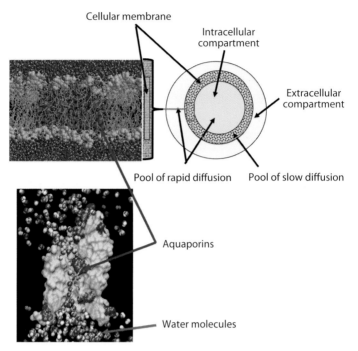

Figure 6.3. Hypothetical model with two pools of water. The slow diffusion pool corresponds to a layer of water molecules interacting with the membranes of cells. Water passage through the membrane can be facilitated by the aquaporins in certain cells.

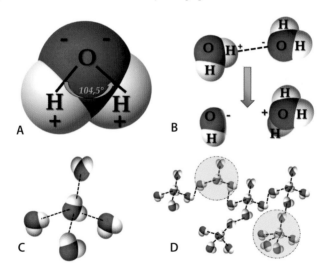

Figure 6.4. A: Water molecule; the 104.5 degree angle between the hydrogen atoms produces an electrical dipole. **B**: Two water molecules sharing a hydrogen atom (hydrogen linking). **C**: "Normal" association of four water molecules. **D**: The network of liquid water includes "defects" (groups of three to five molecules).

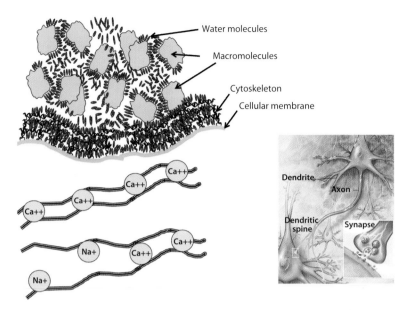

Figure 6.5. The shape of the neuron is maintained by an internal cytoskeleton attached to its membrane. The two threads of the cytoskeleton are held together by calcium ions that carry two positive charges. During activation, the two threads separate, leading to the swelling of the neuron and allowing water molecules to enter.

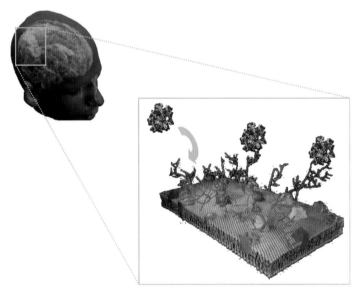

Figure 7.1. Molecular imaging consists of visualizing targets—for example, receptors at the surface of pathological cells—using tracers developed to attach to the target, but also to give a signal detectable through imaging.

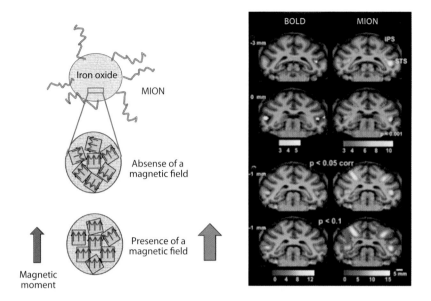

Figure 7.2. A: An iron oxide nanoparticle (MION). In the presence of a magnetic field, the domains that compose the MION are oriented by cooperating. **B**: After injecting a MION (here, in a monkey), the vascular response to the activation is greater than with BOLD fMRI.

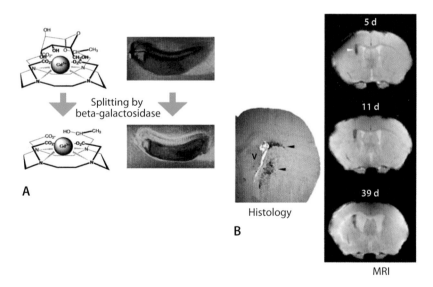

Figure 7.3. A: Following splitting by the galactosidase enzyme in specific cells, water has access to the gadolinium, which causes a contrast to appear. **B**: MRI images after inoculation in the striatum of a rat, of an adenovirus containing a gene provoking the production of an MRI tracer (ferritin) by cells.

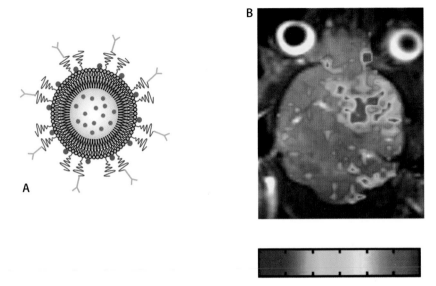

Figure 7.4. A: Emulsion with a base of 19F PFOB. Nanoparticle containing a fluoride compound (derivative of PFOB), producing the MRI signal and covered with complexes to render it furtive and specific of receptors present in tumors. **B**: MRI section of a rat showing the tracer in a tumor.

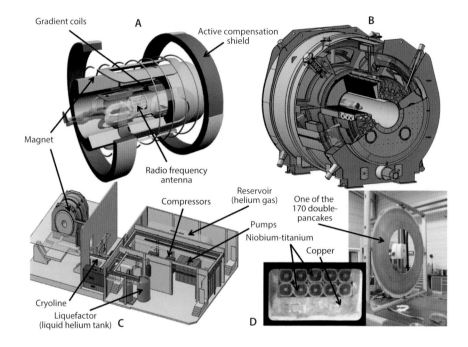

Figure 7.5. Future 11.7 T MRI NeuroSpin scanner (Iseult). **A**: Diagram of the various components. **B**: 3D rendering of the magnet and its cryostat. **C**: Associated cryogenic installation (production of liquid helium at −271°C). **D**: Section of the conductor (10 threads of niobium-titanium) and the assembly in double pancakes.

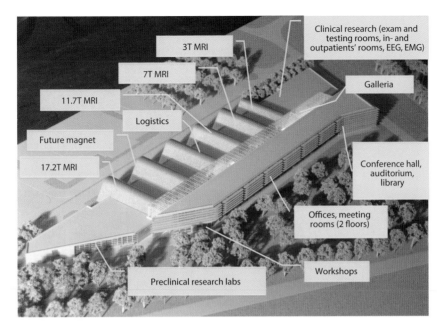

Figure 7.6. Internal organization of NeuroSpin (CEA Center at Saclay).

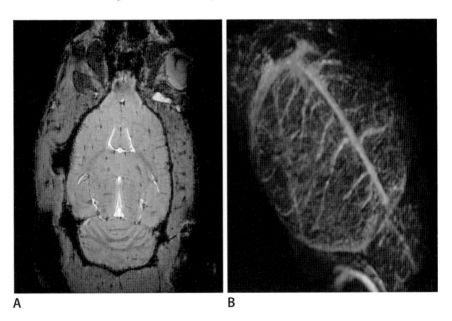

A   B

Figure 7.7. A: MRI section of the brain of a rat obtained at 17.2 T (80 microns of resolution). **B**: 3D rendering of the blood vessels that spontaneously appear visible due to modifications in the oxygenation of the blood induced by anesthesia.

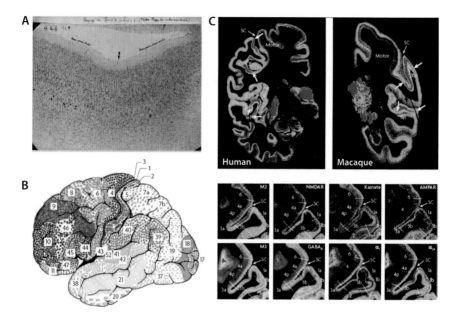

Figure 7.8. A: Microscopic view of the cortex (Brodmann, 1909) showing the structural transition between two adjacent regions. **B**: The Brodmann areas over the entire cerebral cortex. **C**: Maps of receptors showing the various distributions among adjacent cerebral regions.

shows the regions of the body where the FDG has accumulated. And the cells that need the most energy are in fact cancer cells. The PET-FDG images thus show regions where there is cancerous or metastatic tissue.

The PET-FDG method is routinely used, but presents some disadvantages. First, the molecule must be created, made radioactive, and transported, which is costly. In addition, a radioactive injection must be given to the patient. The dose is such that the risk is minimal, but it is all the same an injection that requires particular precautions, in particular for the personnel who do them routinely. Furthermore, the situation is problematic for the detection of tumors in the pelvis and the perineum (prostate, uterine) because radioactive FDG, excreted naturally in the urine, ends up accumulating in the bladder, and can mask a possible tumor in the region. Another limitation comes from the fact that the PET images do not have a very good spatial resolution: fine details (or small tumors) can sometimes remain undetected. This is why the PET camera is often coupled with an x-ray scanner, which enables the PET image (in color) to be superimposed on the very fine image of the scanner. Above all, the PET-FDG method detects all the tissues demanding energy, and those tissues are not all cancerous. This is the case with inflammatory lesions in the vertebrae, for example, following a trauma or a compression that has gone undetected. It is then very difficult to distinguish this inflammatory lesion from a metastasis, whereas the prognosis is of course completely different.

What is water diffusion MRI's role here? As I had already shown at the end of the 1980s, it happens that the diffusion coefficient of water is slowed in cancerous tissue. Since then, this discovery has been confirmed many times, and Professor Taro Takahara in Kanagawa, Japan, in 2004 perfected a simple procedure that amplifies the visible contrast on diffusion MRI images: since lesions have a lower diffusion coefficient, and are thus a priori cancerous lesions, they appear clearly on images (reference 5.9). Since then, the use of diffusion MRI to detect cancer has only grown—in particular, in countries where PET is not very widespread, such as Japan. With diffusion MRI cancerous lesions are visible without the injection of a tracer, the marker being simply the slowing of the Brownian motion of water molecules (figure 5.4). If we add that the spatial resolution of the MRI images is much superior to that of PET, even with the adjunct of a CT scanner, and that there are many more MRI scanners than PET cameras in hospitals and clinics due to the diversity of their use, we can understand why radiologists are increasingly interested in them.

The mechanism through which the diffusion of water is slowed in cancerous tumors is not yet well understood, but it seems to correlate with the rate of cellular proliferation and the density of the cells in the tissue. Since diffusion MRI is quantitative, directly providing at each point of the image the diffusion coefficient of water, we can envision even further progress. A major problem in chemotherapy, very often used to treat cancer, is that it takes a lot of time, often months, to know if the treatment is effective, by observing the clinical improvement of the patient or the shrinking of tumors visible on images obtained via MRI or CT scanners. If the treatment doesn't seem effective, it is sometimes possible to change it, but then one has lost valuable time. Unfortunately, when a cancerous tumor is diagnosed the lifespan of the patient is sometimes short (on the order of an average of 6 months for glioblastomas, cancerous brain tumors). With diffusion MRI, it has been shown that the positive effect of treatment can be visible as of the first days following the administration of chemotherapy: the diffusion coefficient of water rises, is normalized, first at certain points in the tumor, then in all of it, well before the volume of the tumor shrinks or the patient feels better (figure 5.5). If diffusion hasn't normalized after a few days of chemotherapy, we can then deduce that the treatment is ineffective and if possible proceed to a different one (reference 5.10). This revolutionary approach is not yet used routinely, but is currently being evaluated in the United States for brain, breast, and prostate cancers.

Diffusion MRI can also be used for the diagnosis of "noncancers"—in particular, in the breast. In fact, today we know that breast cancer is overdiagnosed. Due to the systematic use of mammography, a number of women find themselves diagnosed with "suspicious" lesions. But these lesions, in particular ductal, in situ lesions, are sometimes benign and will not evolve into cancer. However, these women are in general subjected to surgical intervention, at least for a biopsy whose reliability is not entirely foolproof, which can lead to the removal of the breast (mastectomy) "if there is any doubt." The psychological effect is of course devastating and the economic impact considerable. With my Japanese collaborators we have shown that it would be possible with diffusion MRI to correctly identify those noncancerous lesions and thus prevent those women from undergoing a useless intervention or mutilation (reference 5.11). This advance still must be validated over a large number of patients, but the approach opens very promising perspectives.

MRI could, moreover, be used in the treatment of cancer, and not only for its diagnosis. Hyperthermia, or thermotherapy, consists of heating the cancerous tissue above 43°C to destroy the cells. It is possible to integrate within the MRI imager a system of heating through radiofrequency, as my collaborators and I showed at the NIH at the beginning of the 1990s (reference 5.12). These radio waves are emitted through a series of antennae and concentrate their energy on the region to be treated. What is more, with diffusion MRI it is possible to obtain images of the temperature reached at each point of the heated region. Now, Brownian motion results from the thermal agitation to which molecules are subjected, and the diffusion coefficient depends directly on it, for water, varying from 2.4 percent per degree. We then need only to acquire images of water diffusion continuously during the heating of the tumor, then deduce from the diffusion coefficients obtained from the elevation of temperature at each point, in order to ensure that the healthy tissue is not heated and that the cancerous tissue has indeed reached the temperature required for its destruction (reference 5.13). This idea of obtaining images of heating through diffusion MRI was published, and my colleagues at the NIH, still just as witty, pointed out to me that I had constructed the most expensive thermometer in the world. Another therapeutic approach consists of using focalized ultrasonic networks that "burn" the cancerous tissue for a very short moment. The ultrasonic system is incorporated onto the table on which the patient lies in the MRI scanner. Here, too, imaging of the temperature is important. Several prototypes are currently being evaluated.

MIKE'S OTHER DISCOVERY

But let's go back to the brain. Alongside his discovery of the slowing of diffusion in acute cerebral ischemia, in 1990 Michael Moseley made another observation in his cats, which was to lead to another spectacular and at that time unthinkable application of diffusion MRI. By observing the MRI images obtained from his cats, Mike had noticed that the diffusion was "anisotropic" in white matter, first in the spinal cord, then in the brain. Clearly, the diffusion coefficient was not the same depending on the direction of measurement.

As we saw in chapter 2, white matter is made up of packets of fibers, axons, that emanate from the neurons and enable them to communicate. These sorts of strange "electrical wires" conduct electricity at a speed that depends on the thickness of the fatty insulating casing, the myelin, that envelops them. Mike noted that in a direction parallel to these fibers, the water diffusion was 6 to 8 times higher than in a perpendicular direction (reference 5.14). This discovery was a complete surprise, because up to then all the other parameters to which MRI was sensitive depended on no particular direction. After validating the method over several vegetables of reference (asparagus and other celery that became remarkably prevalent in MRI meetings) known to be made of parallel (vegetal) fibers, thus presenting a priori an anisotropic diffusion, this discovery was quickly confirmed in humans by a few teams, including mine at the NIH.

The explanation is still a bit simplistic (reference 5.15): water molecules have difficulty diffusing perpendicularly to the axons (their diameter is on the order of 1 to 15 microns), since their path remains confined to the interior of the fibers due to the presence of the myelin casing that surrounds them, which is quite certainly impermeable, whereas in the direction of the axons there is no obstacle to oppose diffusion. This hypothesis is probably not very far from reality, but all is still not entirely clear. For example, this asymmetry (the term "anisotropy" is more appropriate) of diffusion also exists (but in a less pronounced way) in the fibers of white matter without myelin, as in newborns.

Whatever the case, it very quickly appeared that diffusion MRI was a unique tool with which to study white matter and its pathologies (such as multiple sclerosis, or MS), with every anomaly in the organization of fibers translated by anomalies in water diffusion. For example, if the fibers are more or less destroyed by an inflammatory phenomenon (the case with MS), the anisotropy is reduced, all the more because the destruction is profound. Since MS is an illness that evolves slowly, it is possible to see which spots become active in order to adapt the treatment. On another register, cerebral development in a baby is intimately linked to the development of connections between the cerebral regions. The functionality of these connections depends on the degree of myelination of the fibers, with myelination occurring until the end of adolescence. Since the degree of anisotropy of the water diffusion is correlated to the thickness of the myelin (and to the organization

of the fibers, with anisotropy being greater in the compact and well organized network of fibers) with diffusion MRI it becomes possible to follow
the development of the brain in a child, in a completely noninvasive way.

WHITE MATTER TAKES ON COLORS

But what "exploded" the diffusion MRI field was the hypothesis I proposed
during my stay at the NIH, by inversing the problem. If water diffusion depends on the orientation of the myelin fibers in space, then it should be possible to determine the orientation of those fibers by measuring the diffusion
in different directions: the direction of the fibers would be that for which the
water diffusion is most rapid. Since diffusion MRI gives the value of diffusion
at each point in space, this approach would give the orientation of the fibers
at each point in the image: in other words, we would obtain an image of the
orientation of the fibers, something that had never been obtained before, nor
had it even been imagined up to then. Philippe Douek, a young radiologist
interning at the NIH, and I thus produced the first images in the world of
the orientation of fibers of white matter in space (figure 5.6; reference 5.16).
The approach was rough because we were measuring the diffusion only in
two perpendicular directions (vertical and horizontal). The direction of the
fibers was determined in function of the relationship of diffusion between
those directions at each point (voxel) of the image: the more rapid the diffusion in the vertical sense, the redder the voxel, and the faster the diffusion
in the horizontal sense, the bluer the voxel. If no difference appeared, the
voxel was green. These images therefore showed the red and blue zones in
the white matter, which corresponded exactly to the expected orientation of
the fibers, vertical or horizontal, in anatomy textbooks! We published this
work, which, for an as yet obscure reason, received little attention (but I was
used to that, the "invention" of diffusion MRI having received the same reception), whereas it is the very foundation of "tractography," the imaging of
intracerebral tracts.

Things progressed rapidly after my encounter with Peter Basser, who
worked in another institute of the NIH. Our encounter during a research
seminar was fortuitous, and decisive. Peter approached me, very interested
because he had worked on ion movements in tissue. He pointed out that
the anisotropy of diffusion should be processed through a formalism with

a base of mathematical "tensors": clearly, the variation of diffusion in space necessitates describing diffusion not as a number (the diffusion coefficient) but as a matrix (the "tensor" of diffusion) that indicates not only the diffusion in each direction but also how those directions are coupled among themselves, in the case when the movements of water following a direction are necessarily accompanied by movements in a perpendicular direction (a bit like on an inclined plane we move both horizontally and vertically). With the tensor the spatial properties of diffusion are described, and it is possible, with mathematical processing, to determine the direction of the space along which diffusion is maximal. Peter's idea didn't surprise the physicist in me, but it remained a serious obstacle: to evaluate each element of the tensor with MRI, which had never been envisioned so far!

After reflecting a bit I was able to resolve this challenge and proposed a new method to Peter. We had to obtain images of diffusion along six fixed directions (by changing the direction of the field gradient) and from this would deduce, from the calculations that Peter had developed, all the elements of the tensor of diffusion at each point of the image (references 5.17 and 5.18). From that tensor another calculation gave the direction of the most rapid diffusion at each point in the event of anisotropy—that is, the direction that was supposed to coincide with the direction of the underlying fibers. The first attempts, achieved with our post-doc James Mattiello, immediately proved positive—we didn't use brains, rather meat (muscle fibers have an anisotropic diffusion due to their elongated form) bought at the corner supermarket. I remember the face of one of my colleagues when I suggested that he take the rest of the rabbit meat that we hadn't used home. He told me his wife would never allow rabbit meat in her refrigerator. And so I learned that in the United States a rabbit is a pet that isn't eaten, but that brings Easter eggs to children. What we found in the supermarkets thus must have been intended for visiting Frenchmen . . .

Very soon we confirmed the success of the method in the brain (first of a monkey). After several attempts it appeared that the best means of showing anisotropy and thus the direction of fibers, was to use a scale of color, red, blue and green, corresponding to the directions X, Y, and Z, and all the composite nuances to intermediary directions (figure 5.6; reference 5.19). But the diffusion tensor also enabled us to determine precisely the degree of anisotropy and the average diffusion (independent of the anisotropic effect), which was of the greatest interest for diagnosing pathology in white matter. Peter, James, and I published the process between 1992 and 1994 under the name of diffusion tensor imaging (DTI). Today installed on almost all of the MRI scanners in the world, it has been remarkably successful: there are more than 200,000 references on *Google Scholar* in 2013.

The reason for this success is that for the first time it became possible to obtain images of intracerebral cabling, a possibility that wasn't even imaginable at the time: even with the brains of cadavers, the paths of white matter fibers are very difficult to follow and dissect, as those fibers are very fragile. In an animal we can see these fibers by injecting an enzyme, peroxidase, which "climbs" into the axons from the injection point. But this method remains invasive (you have to dissect the brain and then observe it under the microscope), and the fibers can be followed only over a few millimeters. With DTI the images of the orientation of fibers are obtained in humans, throughout the whole brain. But these images are only a collection of spots of color, the color depending at each point on the orientation of the white matter (which has then become quite colorful...) (figure 5.7). There was still a step missing in the continuous tracing of the path of the fibers. This step was taken at the end of the 1990s when we borrowed concepts used for artificial vision, the recognition of shapes by computers and robots.

The idea was to determine whether, at a given point (voxel), the orientation of fibers that had been determined there was identical to that of an adjacent, or at least very close, voxel. If this was the case, we could assume that the fibers pass in the two voxels, and that they are connected (references 5.20–5.22). This "tractographic" analysis is then conducted gradually, which enables us to reconstitute, point by point, a hypothetical trajectory of the fibers that have traversed those voxels (figure 5.7). The images obtained (in three dimensions) are spectacular: each fiber is seen by a color and dozens or hundreds of fibers can be represented, a bit like packs of wires that we find in telephone switchboards or on the ceilings of computer rooms. For the first time, we have three-dimensional images of intracerebral cabling (figure 5.8). To achieve this, we had only to lie down in the MRI scanner for a few minutes...

As we can see, the water molecules that have been at issue since the beginning of this book start to play an increasingly important role in the exploration of our brain. For DTI, it is as if these molecules were cars circulating on roads (the fibers of white matter) over which we are flying in a helicopter. However, it's dark, there is fog, and the roads are not clearly visible. The trick is to watch the movement of the cars whose headlights are the lighted, magnetized water molecules, which create an MRI signal. In general, except in the case of snow or sleet, we can assume that the movement of the cars (or their headlights) go in the direction of the roads, and not perpendicularly. This is the equivalent of the anisotropy of the diffusion of water in white matter. We can thus reconstruct the road network from observing the cars' lights. This is the principle of tractography by DTI that enables us to reconstruct the "highways" of the brain in white matter. But the analogy goes much further. Just as through this process the origin and the final destination of the cars and their contents are not known (we observe their movement only over a few meters), tractography does not give us "functional" information about the nature of the traffic within the white matter. The information remains purely anatomical, a simple map of the paths of intracerebral wires, which is still a considerable advancement. Perhaps one day we will also be able to capture the information that is circulating in those fibers.

The DTI method does have some drawbacks. For example, it does not enable us to simultaneously see two or several packets of fibers that cross in a voxel (only one direction can be determined, and it becomes erroneous when two fiber packets cross). And so improvements have been made: this problem of crossing can be resolved when we obtain diffusion images, not from six directions (the minimum to determine all the terms of the tensor), but from twenty, sixty-four, or more. This high angular resolution enables us to determine more precisely the angular distribution of the paths of diffusion of the water molecules. We can then see that fibers coexist in different directions on certain voxels in the image (references 5.23 and 5.24). The images of the intracerebral cabling are truly spectacular and regularly show up on the covers of scientific journals. They have even begun to appear in anatomy textbooks (such as *Gray's Anatomy*, also the name of a popular TV series) used by medical students. These images also reveal that the paths of communication in the brain are not randomly distributed. There appear to be knots, points of convergence, sorts of *hubs* like what one finds on the system-wide maps of airplane companies showing points of connections at airports (reference 5.25). We are beginning to realize that many cerebral pathologies are in fact anomalies of connections—in particular, in certain psychiatric conditions such as schizophrenia.

An Asynchronous Brain

Dr. Thomas Insel, the director of the National Institute of Mental Health at the NIH, in 2012 declared that psychiatric pathologies should be redefined on the basis of anomalies in the cerebral circuits and of interconnections among cerebral regions. It is interesting, moreover, to note that in the United States the medical costs associated with psychiatric illnesses began to be reimbursed by insurance companies in 2008, when legislators started to be convinced through MRI images of the biological origins of these afflictions, and of the possibility of diagnosing them and providing treatment. The "anatomical" MRI images of these patients are most often normal, without visible lesions. Similarly, functional MRI (fMRI) images appear normal. However, these patients have symptoms and suffer. The great majority of schizophrenic patients have auditory hallucinations and "hear voices." fMRI in fact reveals that their auditory cortex is activated when they hear these voices. But, as we saw in the preceding chapter, the simple fact of thinking of images is enough to activate our visual regions. Moreover, we hear our voice constantly, and we speak to ourselves—in particular, when we read "silently." But we know perfectly well that it is our own voice we are hearing.

What has been observed in some schizophrenic patients through diffusion MRI is that the paths of communication between the frontal and temporal lobes are abnormal—in particular, on the left side. We can thus imagine that if there is too great a delay between the internal production of this voice and its perception or reverberation through auditory regions, the patient has the impression that the voice is coming from elsewhere. We also think that the abnormal development of some connections at the beginning of life might contribute to autism. Cerebral connectivity also evolves when neurons degenerate, due to normal or pathological aging, as in Alzheimer's disease or in chronic alcoholism, where DTI has shown that the paths of communication between the two hemispheres (corpus callosum) are altered.

The stakes are high, and the NIH has earmarked $30 million for two groups of American teams to create an atlas of the intracerebral connections in the normal subject, the *Human Connectome Project*. MRI scanners specifically intended for DTI have been developed with systems of very intense field gradients. Similar initiatives exist in Europe (Project CONNECT), although the budgets are lower, and our teams have already obtained quite competitive results. To obtain more functional results, the anatomical connections identified through DTI and its more advanced derivatives are then compared with the functional images obtained "in a state of rest," as we saw in the previous chapter. These images provide an idea of the networks of regions that constantly "speak to each other," and are thus a priori connected via paths of communication identified by diffusion MRI. DTI imaging, moreover, also enables us to label the cerebral regions in function of their connections, and to correlate them with the regions identified from images of cerebral activation. For this we start with central cerebral regions, such as the thalamus or the basal ganglia (see figure 1.1), and we determine the connections that come out of them. Thus we rediscover the subregions of the thalamus that are connected and are projected along very specific regions toward the cerebral cortex.

An interesting question, which has not yet been fully addressed within the scientific community, is how these maps of connections would enable us to study synchronization in the brain. The study of the synchronizations between cerebral regions is already quite old, resting principally on the recording of electrical (EEG) or magnetic (magnetoencephalography, or MEG) signals produced naturally by the brain. A given region of the brain is constantly bombarded with a great deal of data coming from other cerebral regions. It carries out a process that is unique to it (and that remains still to be understood), and a resulting signal is emitted toward other more or less distant regions. All the same, the information along the axons does not circulate at the speed of light, but at a few dozen meters per second in the myelinated fibers, much less in others, so that the axons can be considered as "lines of delay." It thus takes some hundredths or thousandths of seconds for the information to go from one end of the brain to the other, even more if this information is being relayed in certain structures such as the basal ganglia at the center of the brain. This in part explains the length of our reaction time. The information that arrives at a given region at a given moment coming from other cerebral regions has not been emitted at the same moment. It is a bit like our vision at night under a starry sky: the light of the stars that reaches us has been emitted by those stars at quite variable moments depending on their distance from the Earth, their light circulating at the speed of light, which is certainly very rapid, but not infinite considering the dimensions of the universe. The light of certain stars may have taken 20 million years to reach us, 40 or 50 million for others that are, however, located apparently close to the celestial vault. The "instant" sky that we see thus does not correspond to a physical reality, and we must use huge calculators to reconstitute the real sky at a given moment.

How does the brain account for these temporal delays between regions to process information? It also uses calculations to anticipate, project us into the future, and make us react "unconsciously" in an appropriate way: if we slip on a bar of soap in our bathtub, our brain knows it is better to hang onto a nearby bar rather than the shower curtain . . . Similarly, the brain of a tennis player, from visual information about the ball that is arriving, calculates the position for the body and the hand (which necessitates the controlled activation of a great many muscles) so that the racquet encounters the trajectory of the ball. Thus the view of the world our brain gives us through our eyes is not real. We know that it is an anticipation, a "prediction" of what could be reality in 100 thousands of a second. In other words, our brain will give us a view of the end of the world before it happens (but we won't have much time to take shelter . . .)!

We can imagine that any anomaly in the time it takes for information to travel between cerebral regions can seriously disturb brain function, just as a late train can block an entire regional network. It is therefore not surprising that anomalies in connections might cause the problems observed in schizophrenia or autism (the origins, genetic or acquired, of these anomalies remain to be clarified). After all, we feel very uncomfortable when we watch a film whose sound and image is out of sync, and if the delay is too great, the meaning may be completely lost.

Water diffusion MRI thus offers considerable perspectives for an understanding of the mechanisms of intracerebral synchronization, enabling us to identify the paths of communication, and potentially to calculate the time of transit along the paths. This is particularly important for an understanding of language, for we must be able to identify sounds that are sometimes very similar, which may differ only for a very brief moment. For example, the perceived difference between the sounds "ga" and "ba" occurs in the first 20 thousandths of a second when those sounds are heard. Several studies have already shown how diffusion MRI and DTI might be used to identify the anomalies in connections. In dyslexics, who have difficulty learning to read, the anisotropy of the water diffusion in the left temporoparietal regions (linked to hearing and language) is lower than normal, which suggests that the connection fibers are badly organized or less effective, and even more so if the dyslexia is severe (reference 5.26). These patients also often have difficulties recognizing voices. An improvement following therapy seen in some dyslexic patients is accompanied by a partial normalization of the anisotropy of water diffusion—in other words, of the reorganization of connection fibers that become more functional—which is extremely encouraging given the possibilities offered by cerebral plasticity.

The paths of communication between cerebral regions are present very early in life. DTI images obtained in 2-month-old babies show that most of the major paths of connection that we find in adults are already present (references 2.1, 5.27, 5.28). The anisotropy of diffusion remains weak, all the same, because in general these fibers are not myelinated, and are thus nonfunctional. The myelination process takes time (at least until the end of adolescence), and the primary or cognitive functions appear in the course of myelination. It thus takes several months for the visual regions to be functionally connected and for the baby to be able to fully "exploit" what it sees. DTI imaging has been able to objectify this myelination process (an increase in the anisotropy of water diffusion) and, for the first time, to show that it begins first in the extremities, and is not homogeneous along the network of fibers. For language, myelination takes several years, but the specialization of the left hemisphere appears very early on, well before the acquisition of spoken language. Just as fMRI reveals that in a 2-month-old baby the left temporal regions are more activated by speech than the right, DTI imaging shows that certain frontotemporal connections (arcus fasciculus) are already, between 1 and 3 months, more developed on the left than the right. DTI imaging is beginning to be used to evaluate cerebral maturation in babies and preemies when anomalies are suspected (figure 5.9).

DTI imaging also enables us to do "neuroarcheology," to dig into our brain to discover how it has evolved in the course of our life. Thus the study of the brains of adult professional pianists 30 years old reveals traces of how they learned their instrument, by correlating the degree of anisotropy of diffusion with the number of hours of practice spent during different stages of their lives. The anisotropy of diffusion (and thus the structuring of fibers, density of axons, myelination, and so on) in the "internal capsule" (where the fibers of motor commands pass), and in the isthmus of their "corpus callosum" (structure of white matter that connects the two hemispheres and plays a fundamental role in two-handed coordination) (see figure 1.6), is directly linked to the number of hours of practice before the age of 11 (between 1,000 and 3,000 hours) (figure 5.10). In comparison, these regions don't evolve noticeably in a controlled population of nonpianists. Another area of the corpus callosum necessitates the increased hours of practice required to advance between the ages of 12 and 16, while the arcus fasciculus that connects the frontal lobe to the parietal lobe necessitates between 15,000 and 35,000 hours of practice beyond 17 years old to be at the level of a professional pianist (reference 5.29). These studies show how training on the instrument in a certain sense sculpts the brain, but they also stress the importance of learning when we are young, as the brain is modeled much more quickly and much easier at an early age, because when we are older it takes around ten times as many hours to reach a comparable quantifiable change in the structuring of the white matter.

After looking at white matter, and the connections that it establishes between cerebral regions, as revealed by the anisotropy of water diffusion, it is time to return to gray matter, to the realm of neurons and our thoughts. What will water molecules reveal about the functioning of our brain?

SIX

WATER: MOLECULE OF THE MIND?

"If there is magic on this planet, it is contained in water," Loren Eisely, the American philosopher, anthropologist, and essayist wrote in 1946. Up to now, this book has largely been a story of water, and it will remain so, but whether we're talking about water made radioactive for PET, or "magnetized" water for MRI, it plays the role only of passive vector, although this intermediary position between physics and biology is of immense importance for such a small molecule. But mightn't water play an even more important role, as it is implicated in the very functioning of our brain?

Eighty percent of our brain weight is made up of water (nine molecules out of ten are water molecules), which is certainly an indication of its importance. However, do we really know what water is? Perhaps, we know that it is H_2O, two hydrogen atoms and one oxygen atom, but D. H. Lawrence, the early twentieth-century English writer, said that there is something else that makes it water, but we don't know what it is . . . which is still somewhat true in 2014. But before going any further into the mysteries of this very strange molecule, let's look at the way in which water and its eternal Brownian motion enable us to see our brain at work.

THE FIREFIGHTERS ARRIVE A BIT LATE

We have learned in the preceding chapters how it is possible to see the brain think with BOLD fMRI via a variation in blood flow (and in the state of oxygenation of the associated hemoglobin) that is produced in activated regions.

This approach has great potential, and thousands of scientific articles have been published showing the activation of cerebral regions during a number of sensory-motor and cognitive tasks, as seen in the often impressive examples in chapter 4. Yet the method does have some limitations. The first is that it is relatively slow. In fact it takes a few seconds (around 6) after the beginning of neuronal activation for the change in blood flow to reach a peak (figure 4.1; reference 6.1). Now, it is clear that our brain works more rapidly than that; otherwise we would be run over crossing the road . . . The second limitation is its spatial imprecision: the blood vessels irrigate (arterial) or drain (venal) the territories whose size is on the order of a few millimeters (reference 6.2), or well beyond the functional neuronal units that occupy around a fraction of a millimeter. Thus we cannot determine a priori which of these units is involved when a region is seen as "activated" on BOLD images.

The situation is a bit like that of a neighborhood in which a house is on fire. What is detected by the BOLD method is in a certain sense the increase in the flow from the fire hydrant at the entrance to the neighborhood, and not the house that is burning. And the analogy goes even further: the flow from the hydrant only increases after the arrival of the firefighters, or well after the beginning of the fire, and it will continue to remain high a bit after the extinction of the fire, out of precaution. If the firefighters don't come, the house will burn down, but our system, based on the flow of the hydrant, won't alert us to that. And another situation also occurs with the BOLD method: in the presence of certain drugs (anti-inflammatories, antihistamines taken for seasonal allergy rhinitis), the BOLD response is inhibited, and even completely abolished in certain pathologies such as arteriovenous malformations that maintain the blood flow permanently at a high level. The flow can henceforth not increase (to resume our analogy, the firefighters are busy elsewhere and are already using the fire hydrant), and the neurovascular coupling is defective. It would be particularly unfortunate in testing certain patients not to be able to see any activation with BOLD fMRI near such malformations, and to conclude that the surrounding brain tissue isn't functioning and could be taken out with the malformation! In fact, it probably functions completely normally.

We can see, then, that even if the BOLD method has many positive uses, it would be best to have a more direct method to detect neuronal activity.

But for this it is essential to escape the sacrosanct dogma of neurovascular coupling, which is the principle upon which functional neuroimaging has been based for 40 years. And this is no small matter, not only on the scientific and technical levels, but also on the level of simple practicality, for as we have seen, conservatism often reigns in the scientific milieu.

It was within this context that, between 2000 and 2006, I put into place the foundations of a completely new approach to obtain images of cerebral function from images of water diffusion. The idea was as follows: for around 50 years researchers had noticed that changes appeared in the microscopic structure of cerebral tissue when it is activated (reference 6.3). These changes were detected, directly or indirectly, from optical procedures on animal tissue, simple "living" slices of the brains of rodents, or in situ in animal (primate) brains. By sending a beam of light onto the cerebral tissue, the researchers saw that the photons of light had deviated in the tissue, the cells performing as obstacles to the photons, and that above all this deviation changed temporarily when the tissue was activated.

The physical structure of the tissue thus changes during its activation, and the underlying hypothesis is that the cells involved (neurons, some of their parts, and perhaps also glial cells) swell when they are activated. This neuronal swelling has been verified by observing the movements of the cellular membrane of nerve fibers through piezoelectric captors (sensitive to pressure). These experimental facts were known by the neuroscientific community, but not by the world of MRI physicists. What the physicists knew was that water diffusion is slowed in cerebral ischemia (stroke) in the acute phase, as we saw in the preceding chapter, and that this diffusional slowing is correlated with cellular swelling linked to the ischemia.

SWELLING NEURONS

Other groups have seen similar results. For example, the stimulation of a point on the cerebral cortex of a rat through potassium ions leads to a hyperreaction (similar to epilepsy) that is propagated along the cerebral cortex at a speed of several millimeters per minute (a phenomenon called "spreading depression") (references 6.4 and 6.5). Images of water diffusion acquired simultaneously revealed that diffusion slowed in the stimulated region, the slowing then propagating along the cortex in the same way. This phenomenon has been associated with underlying cellular swelling.

Other studies confirmed that when cells swell—for example, when they are placed in a solution very weak in ions and solutes—the diffusion coefficient of water measured by MRI diminishes. All the same, the exact mechanism linking cellular swelling and the slowing of water diffusion still remained a mystery. And so I asked the following question: would it be possible to observe, in humans, and in conditions of physiological activation, a slowing of water diffusion in activated cerebral regions? The physiological swelling responsible for the phenomenon would be much smaller than the pathological one observed in cerebral ischemia or "spreading depression," but perhaps it would be enough, knowing that a large number of neurons are activated at the same time in a given region, and can be detected through diffusion MRI.

After a positive preliminary study published in 2001 in *Proceedings of the National Academy of Sciences* (*PNAS*) (reference 6.6), I continued to work on this idea with my Japanese collaborators in Kyoto, in 2005. The results were spectacular. The experiment consisted of showing volunteers a blinking checkerboard for 10 seconds at repeated intervals. This was the type of stimulation that had been used by the precursors of fMRI twenty years earlier, as this stimulus strongly activates the visual cortex. But this time the images acquired during the stimulation were images of water diffusion. Not only did the water diffusion diminish, slow, in the visual cortex each time the checkerboard was shown to the volunteers, but the slowing began exactly at the start of the stimulus—that is, up to 3 seconds before the start of the BOLD (vascular) response—and stopped exactly at the end of the stimulus, whereas the BOLD response still continued to climb for a few seconds. The images of activation were even much clearer with diffusion MRI, and the activation corresponded exactly to the visual cortex, whereas the BOLD activation, due to its vascular origin, was more diffuse (figure 6.2).

For us, the temporal lag between the diffusion images and the BOLD images proved that we were dealing with a phenomenon other than the classic vascular response, although we could not assert that it involved neuronal swelling. In any event, this diffusional slowing "jibed" much better with the temporal waning of neuronal activation, and had most certainly to correspond to a more direct effect than the BOLD vascular response. The results were published again in *PNAS* (reference 6.7), but the reception by the community of MRI physicists was cool, and it continues to be. On the contrary, the neuroscientific community considers it a promising approach.

Some groups, noting other experiments, suggested that our results were "artifacts," as the signal obtained in diffusion MRI was contaminated by the BOLD signal (reference 6.8). Since then we have published several articles showing, both theoretically and experimentally, that this contamination, although real, does not explain our results and that there is indeed a diffusional slowing in activated cerebral regions (references 6.9 and 6.10) well beyond the visual cortex, in tests of language or memory, for example, and that this slowing is always largely in advance of the vascular response. What is more, a group in Florida has shown (on living, bloodless sections of rat brain deprived of all vascularization) that an activation by a neuromediator provokes a diffusional slowing visible in diffusion MRI, associated with neuronal swelling (reference 6.11). In spite of that, another group of colleagues published an article confirming our results (but much less clearly), while concluding that they "did not believe" in an origin other than a vascular phenomenon, even if they had no other explanation to propose (reference 6.12). In response we lately published another article in PNAS showing that when the neurovascular coupling mechanism was abolished in a series of rats using a drug called nitroprussiate, the BOLD fMRI response vanished, as expected, while the diffusion slowdown persisted. Will this study finally convince our skeptical colleagues?

Thus goes science, and only time and repeated studies can change the reality. I had encountered the same incredulity in 1985 while presenting the premises and the first results of diffusion MRI, and it took close to 10 years for the method to be routinely used; the same thing happened with the invention of DTI (1992–1994) and its routine use (middle of the 2000s). These experiences perhaps explain why the journal *NeuroImage* asked me to write an article on the 25 years of the history of diffusion MRI (reference 4.13). With patience and perseverance, everything ultimately happened. It took close to 20 years for the method I had invented and published in 1988 in the American journal *Radiology* (the primary reference in medical and scientific publications in the field of radiology), which obtained images of microcirculation (perfusion) from images of diffusion (a process known by the name of intravoxel incoherent motion, or IVIM) (reference 5.5), to leave its phase of hibernation. This article had, however, been accompanied at the time by an extremely laudatory editorial (reference 6.14). The method is currently experiencing a remarkable popularity, notably for the diagnosis of cancer (breast, prostate, uterus, liver, and so on), and is the subject of entire sessions at meetings and of many articles . . . more than 20 years after the initial article of 1988. And so, in 2008 I was the one that Herbert Kressel, the editor-in-chief of the journal, asked to write an editorial on the subject, which I titled "IVIM Imaging: A Wake-up Call" (reference 6.15).

Two Types of Water

While waiting for the community of MRI physicists to be convinced of the potential of fMRI through water diffusion imaging, I made progress on a fundamental question revolving around almost everything we have seen in this chapter and the preceding one: how can we explain that water diffusion, a purely physical and molecular phenomenon, is so sensitive to what is happening in the tissue, and in particular to cellular swelling? It was indeed this idea that was at the origin of the work of 1985 that enabled me to invent and develop diffusion MRI, that sensitivity having largely gone beyond my hopes, as shown in the examples of the preceding chapter. In 2002 I organized a colloquium in Saint-Malo, France, with experts from around the world, in an attempt to understand the mechanisms that underlie the modulation in water diffusion in tissue. The colloquium was enthralling, both scientifically and socially, but nothing tangible came out of it. My work, as well as that of other groups (reference 6.16), suggested that there are two types of water in tissue (in the brain as well as other organs), water with relatively rapid diffusion (2 to 3 times less than diffusion in a glass of water at 37°C), and water with relatively slow diffusion (10 to 12 times less than regular water). The existence of these two "pools" of water in biological tissue is also the object of controversy, but the "two pool" concept is rather widespread.

We were able to show during neuronal activation of the visual cortex in humans (which others showed in rats) that water diffusion in rapid and slow pools doesn't change, which is rather reassuring from the point of view of physics (references 6.7 and 6.11). What does change is the relationship between these pools: a slowing of diffusion during activation or neuronal swelling results from an inflation of the slow pool to the detriment of the rapid pool. A fraction of water molecules (1.7 percent in the case of visual activation) thus passes from the rapid pool to the slow pool, which gives the illusion overall that diffusion in the activated region has slowed.

The physical origin of these two types of water remains unclear. It is often suggested that the pool of rapid diffusion might correspond to water in extracellular space, while intracellular water, more confined inside cells, would be at the origin of the slow diffusion pool. In the case of cellular swelling, we can then imagine that the quantity of slow diffusion water increases. Unfortunately, this naïve schema is contradicted by several facts, first the relative fractions of these pools: the rapid pool contributes to around 70 percent of the diffusion MRI signal, whereas the extracellular space is less than 20 percent of the volume of the tissue. In fact, it increasingly appears that these pools do not perhaps have direct physical significance, but might rather have a functional origin.

Water molecules, during diffusion, pass from one pool to the other, and it is the fraction of time that they respectively remain in these pools that would be the important parameter. The structure that is most likely responsible for the modulation of the diffusion process in tissue through physiological or pathological conditions would thus be the cell membrane. In the nineteenth century, it was discovered that the content of cells is "gelatinous"; the notion that cells are surrounded by a membrane was introduced later to explain the absence of mixing between the cellular content and the exterior environment in which the cells are immersed. Today, we know that this membrane, whose thickness is less than a thousandth of a millimeter, is made up of two layers of fat (phospholipids) in which proteins literally "float." At first the membrane was considered to be semipermeable because it was observed that it allowed water to pass through, but not other molecules. But it was ultimately discovered that some ions (potassium, sodium, and so on) could also pass through the membrane in certain conditions. The environment outside the cell is "salty," rich in sodium because life came from the sea, and we retain the vestiges of that legacy. Potassium, however, is confined within the cell.

To account for the passing through of ions, the notion of "channels," sorts of "keyholes" made of proteins through which ions or molecules selectively pass, was invented in 1941. A very large number of these "channels" were identified, and the membrane began to resemble a sieve . . . except that not everything that wanted to could pass through. Ions approach the membrane through diffusion, but diffusion being a somewhat slow phenomenon, nature endowed the channels with pumps that accelerate the exchanges between the interior and the exterior of the cell (figure 6.3).

There are a large number of channels and pumps to accommodate many sorts of molecules, ions, sugars, and amino acids, and these pumps consume a lot of energy. The surface of the cell is thus an extremely complex structure. We also find on it specific receptors that signal to targeted molecules of the cell that other molecules they have recognized, beneficial or harmful, are in the vicinity so that the cell can take appropriate measures to react. They can use those signals for their own needs, or on the contrary, to defend themselves (immunity), even sometimes, if it is necessary for the "collectivity" of the tissue to which they belong, to unleash a "suicide" mission (which cancer cells refuse to do). The membrane is thus the "skin" of the cell, its interface with the outside world. And just as our skin covers the muscles and skeleton that give shape to our body, the membrane is attached to a cellular skeleton, the "cytoskeleton," which is made up of fibers and maintains the general architecture and the shape of the cell. To sum up, the cell, with its envelope and its contents, is far from being a simple structure: a lot of things go on in it, and it would be wrong to consider that the intracellular environment is just a "soup."

104.5 DEGREES: THE ANGLE OF LIFE

As for water, it is probably the most fascinating molecule in the world. "The structure of the water molecule is the essence of all life," wrote the Nobel Laureate Albert Szent-Gyorgyi in 1937. And my high school physics teacher, M. Gendreau, had struck me by telling us that: "We owe our lives to the 104.5 degree angle between the two hydrogen atoms and the oxygen atom of the water molecule." In the end, we exist on very little . . . If that angle were a bit larger (109.5 degrees, for example, as for the carbon atoms of diamonds), we would be quite stiff; and if it were smaller, we wouldn't be here to discuss it. The water molecule has a geometry close to that of a tetrahedron and a pentagon, without being either, which makes all the difference.

In 1781 the Englishman Henry Cavendish discovered the structure of the water molecule with its two hydrogen atoms and its one oxygen atom. However, it is only today, thanks to numerical simulations on extremely powerful

111

computers, that we are beginning to know a bit more about what makes this molecule so special. Due to the 104.5 degree angle, the two positive charges carried by the two hydrogen atoms and the two negative charges carried by the oxygen atom are located on the two extremities, and the water molecule becomes a strong electrical "dipole" (figure 6.4). Water molecules thus attract each other, from the hydrogen atom of the one to the hydrogen atom of another, forming the "hydrogen bond" that was discovered in 1920. The hydrogen bond is a truly passionate link, because the hydrogen atom that has been infatuated with a neighboring water molecule is incorporated into it, and the molecule then temporarily has three hydrogen atoms for itself alone . . . before casting one of its hydrogen atoms off onto another water molecule, and so on.

The entire process is dynamic, with the exchanges occurring in a few billionths of a billionth of a second. Since water molecules can "lose" two hydrogen atoms and "win" two by way of their oxygen atom, they are on average regrouped in fours, and the network of water, whether in a glass or in our cells, is tetrahedral. All the same, this arrangement is not perfect in liquid water (otherwise the water would be a solid crystal), and locally there are many defects in the network. Although the motif of four water molecules is the most common in "linear" hydrogen bonds, the water molecule sometimes has a "forked tongue" with bifurcated liaisons, which creates arrangements of three or five, or even two or six molecules (figure 6.4). This variability prevents water from being structured over a great distance and enables it to remain in a liquid state at the ambient temperature (reference 6.17). Other schemes have been proposed (for instance, by Jerry Pollack) based on the stacking of hexagonal units sharing some hydrogen atoms: in such ice-like networks water behaves more like H_3O_2 (with a net negative charge) than H_2O! The study of the structure of liquid water has always being a very active research topic, confirming that the secret of water is hidden very well in its social behavior.

The consequences of hydrogen bonds are impressive. First, it is this that explains why water is liquid in temperatures that are found on the Earth. Indeed, through its molecular size, water should be a gas at those temperatures, like carbon gas CO_2 which also has three atoms and is about the same size, but does not have hydrogen bonds, which enables CO_2 molecules to wander independently at a distance from each other. But water molecules cannot escape very far and they form a dense network: a liquid. If carbon gas is cooled enough, it also becomes liquid, then solid ("carbonic snow"), even more compact when it falls to the bottom of liquid CO_2. Because of its bi-polar structure and its "magical" angle, the water molecule also distinguishes itself here: ice is formed at 0°C, but is lighter (and thus less dense) than liquid water (the highest density of water is at 4°C). Thus it floats, for which the passengers of the *Titanic*, unfortunately, paid the price. If the water molecule behaved like other molecules, there wouldn't be any icebergs and ice cubes would not float in a glass of whisky.

Hydrogen bonds also intervene in the formation of clouds. Air contains a lot of water, but in the form of invisible vapor, as the molecules are very dispersed. If the density of the molecules goes beyond the capacity of the air to contain them (in function of the temperature, and especially in the presence of charged particles), the water molecules will clump together, while attracting others: a tiny drop (now visible) is formed, then others, and finally a cloud. If the drops are large and heavy enough, they fall in the form of rain. Without a doubt, life would not be possible without such hydrogen bonds. If its intensity were a bit weaker, water would begin to boil at 37°C, the temperature of our bodies, and if were a bit higher, our brain would change into a block of ice. Because of those bonds, water possesses remarkable thermal properties. It has a very large "calorific capacity": a classic magic trick consists of putting water in an (impermeable) paper filter and placing it all on a flame. The paper doesn't burn, the heat being absorbed by the water that begins to boil, keeping the temperature at 100°C. In contrast, water has a high "latent heat" of evaporation: to go from liquid water to vapor, it takes a lot of energy to break the hydrogen bonds, which notably makes the temperature of our bodies lower and cools us when we are sweating. Our ability to maintain the temperature of our body is intricately bound to the water inside it, and to its exchanges with the exterior on the level of the skin. This is why babies, in whom the body surface/volume ratio is much greater than in an adult, are very sensitive to variations in hydration (hyperthermia in the case of fever, dehydration through diarrhea, and so on).

The hydrogen bond, which might appear weak, is at the origin of considerable forces. Some Japanese scientists had fun "peeling" the membranes from cells. Contrary to what was expected, the cells remained intact for hours. Unlike water in a broken flask, the watery, gelatinous content of the cell didn't flow out, but remained whole. How is that possible? It is because the proteins present are large molecules carrying many negative electrical charges on their surface. Water molecules, also charged, attach to each other via hydrogen bonds and with the amino acids of the proteins. The network of water molecules present in the cell represents a considerable force of cohesion, maintaining the structures—proteins, enzymes, DNA, and other cytoskeletons—in place: these macromolecules are entirely shielded by one or two layers of water molecules, attracted by the electrical charges on their surface. In fact, the geometrical shape of the proteins, which is directly linked to their function, is determined by the layers of water molecules that cover them.

PROTONS PLAY LEAPFROG

Now that we are more familiar with the social life of water molecules, it is time to return to the subject that concerns us—that is, the diffusion of the water molecule, and its impact in our brain. In their dense molecular network, we can imagine that water molecules have trouble moving rapidly. In fact, oxygen atoms, to move, must rid themselves of their hydrogen bonds with two water molecules at the same time, which would take a lot of energy and time, much more than the 7 billionths of a billionth of a second that it takes a water molecule to move the length of a molecule. But the nuclei of hydrogen spend their time flirting between one molecule and another, forming temporary "water" molecules with three hydrogen atoms. The hydrogen nucleus, which is the particle (proton) that creates the MRI signal, thus travels much more quickly by jumping from one molecule to another (1 to 2 billionths of a billionth of a second), following the "Grotthuss mechanism," after Theodor Grotthuss, who described it in 1806. In spite of that, water diffusion, observed experimentally, is still too fast to correspond to this mechanism. Recent simulations using super calculators, calling upon quantum mechanics and the "tunnel" effect, suggest that it is the presence of defects in the network of water molecules that enables this rapidity of diffusion, the hydrogen nuclei taking advantage of the confusion to belong to two molecules at the same time, with a lucky oxygen atom being able to find itself for a (very, very) short instant with five hydrogen atoms, then finally nine by "sharing" with three other neighboring water molecules (reference 6.18).

It is this abnormal complex, this defect in the network, that gradually propagates at a rate equivalent to almost two molecules at the same time, explaining the abnormally fast diffusion rate of water, unlike other molecules. On the other hand, an increase in the structure of the network is translated by a slowing of diffusion in the network due to a decrease in the number of defects that result from it and a decrease in the network density. Those very basic phenomena could, indeed, support life as we know it.

The Tribulations of Water in Cells

With all these elements in mind, let's now try to understand the origin of the two pools of water identified by diffusion MRI, and what happens during cellular swelling, whether through neuronal activation (physiological) or cerebral ischemia (pathological). Water molecules in cells absolutely do not differ, of course, from those that are in the oceans or in our bathtubs. What changes is the environment, as found in the bonds within their network. In a cell, one principally finds macromolecules covered with water molecules and exchanging hydrogen atoms with them via hydrogen bonds. All of this is very rapid, but slows the movement of water diffusion, as the molecules remain stuck to these proteins for a short moment. Once unstuck, these molecules must move by shifting around the macromolecules, which slows them down even more. It is a bit like a buyer in a crowded store: his movement is slowed because he stops from time to time in a department that he's interested in, but also because there are a lot of people in the store and moving from one department to the other isn't smooth. Thus, we can assume that the average water diffusion in the cell is around 50 percent slower than in a simple glass of water; yet this cellular water corresponds a priori to the so-called rapid pool.

The slow pool would correspond to molecules that have interacted with other structures that slow them down even more: membranes. We have mentioned the membrane that covers the cell, but we must remember that within the cell there are also other small elements that are also covered in membranes, vesicles, "mitochondria," and so on. The water molecules that are in the presence of these membranes, mitochondrial or cellular, have three mechanisms as they pursue their path: they can retrace their steps ("rebound"), go through walls, or remain stuck before the obstacle (see figure 5.2).

115

One mechanism is responsible for "restricted" diffusion: the water molecules are free to diffuse into the cell, but cannot escape it. But if this happens to some of them, it seems that most of the water molecules in fact manage very well to pass through the membrane, except in special cases. One of these "cases" would involve the axon of the neuron surrounded by myelin: the lipidic (fatty) casing surrounding the membrane is in fact impermeable to water molecules. All the same, this insulating casing is interrupted at regular intervals ("Ranvier nodes") and it is not certain that water doesn't in fact manage to escape. The difficulty in escaping a cell is quite probably at the origin of the anisotropic effect in white matter that we discussed in the preceding chapter.

Most water molecules thus pass through the cell membrane, which is largely permeable, but they are slowed down considerably. The extracellular space is very thin because cells are most often joined end-to-end, representing a network of pores whose bond has a width of around 150 to 200 water molecules on average (the extracellular space contains only 10 to 20 percent of water in the cerebral tissue). Water molecules that have "escaped" a cell thus very quickly find themselves facing another membrane where they can again remain "stuck," or that they will circumvent, or penetrate.

A more effective mechanism (but nevertheless a source of slowing) for passing through a membrane is resorting to the membranal channels mentioned earlier. Some water molecules try to take advantage of the channels made for others . . . Most ions, sodium, potassium, and so on, are hydrated— that is, surrounded by a few water molecules owing to the electrical charges that attract them. They thus move around accompanied by their burden. All the same, the proteins of the channels (like those of the "potassium" channel) often begin by getting the ions that they pass through rid of their water. A notable fact, the movement of water through a membrane is above all due to its own channels: "aquaporins." The aquaporin was discovered by chance on red blood cells, then found on the membrane of many cells, which earned the American Peter Agre the Nobel Prize in Chemistry in 2003. Since Agre's discovery, thirteen sorts of aquaporins have been discovered, present on membranes of almost all types of cells in an organism. In the brain, we principally find aquaporin 4. This protein is found on the membrane of glial cells, and controls the movement of water in those cells. The aquaporin is a very effective molecule that can allow a large number of water molecules to pass every second by making them pass one at a time, in order to prevent only hydrogen nuclei from passing through (figure 6.3). The aquaporin reorients each water molecule individually so that it doesn't establish a hydrogen bond! Depending on the number of water molecules that enter or leave, the volume of the cell varies; the aquaporins are the guardians of the shape of the cell, making it swell or shrink.

We now know that some antiepileptic or migraine drugs act by blocking the aquaporins of glial cells, which highlights their importance in cerebral function. Notably, however, no aquaporin has yet to be found on the membranes of neurons. Is it a matter of time before we discover a specific aquaporin of neurons, or do neurons have another mechanism to control the flow of water through their membrane? We don't yet have the answer to this question, but it is essential to understand the changes in water diffusion that we have observed during cerebral activation.

Another mechanism likely to slow water diffusion through membranes is connected to the structure of water itself. Membranes are covered with alternating electrical charges, positive and negative, on their surface. Water molecular networks can thus find themselves attracted. Since the surface of membranes present planar structures, in contrast to proteins that have the shape of a three-dimensional heap, water molecular layers "fixed" to membranes in turn create a cover of charges capable of attracting other water molecular layers. Gradually, a cloak of several layers of water molecules can thus stack, covering membranes with a structured water mantle. The network achieved by these water molecules is very structured because the position of water molecules is dictated by the charges of the membrane and within the network. And so water diffusion in such a structured, ice-like network is much less rapid than in a normal network presenting many defects. Thus, we should observe a layer of slow water diffusion surrounding the cellular membrane (figure 6.5).

This at least is the thesis I proposed in 2007 to explain, in part, the pool of slow diffusion water revealed by diffusion MRI (reference 6.19). This thesis has not (yet?) been universally accepted, because there is a controversy about the thickness of this layer of water. Some physicists maintain that the layer cannot exceed two or three thicknesses of water molecules, whereas some biophysicists estimate the layer to be hundreds of water molecules (references 6.20, 6.21). Our experimental data suggest a thickness of only a hundred water molecules on either side of the membrane, which, in passing, stresses the extreme sensitivity of diffusion MRI. Also attached to the membrane are other charged molecules, the "glycocalyx" outside a cell, a sort of coat of mail that protects the cell and enables it to be recognized by its counterparts, and above all, in the interior, the cytoskeleton, a network of relatively rigid fibers that give it its shape. These structures also polarize, "structure" the water molecules and might thus contribute to the existence of that layer of slow diffusion water around the cellular membrane.

The membrane thus does all it can to slow water diffusion in its environment, and is quite certainly at the origin of the variations in water diffusion revealed through MRI in various physiological and pathological conditions. Thus, cellular swelling, weak during cerebral activation or greater in acute cerebral ischemia, is accompanied by an increase in the membrane surface and thus by an amplification of the number of slowed molecules. This inflation of the pool of slow diffusion explains the overall lowering of the diffusion coefficient observed in MRI in these conditions, as we have been able to observe very clearly (an increase in the diameter of the cell by 5 percent causes its surface to grow by 10 percent). Similarly, cellular proliferation that characterizes cancerous tissue is translated by an increase in the density of the membranes in the tissue and by an apparent lowering of diffusion. If treatment is administered, the death of the cells and the destruction of the membranes are seen in an increase in the diffusion coefficient, a sign that the treatment is effective. Thus diffusion MRI can be used primarily as a marker of the status of cellular membranes in tissue. Similarly, in the case of the anisotropy of diffusion in axons, we can deduce that the diffusional slowing observed perpendicularly to the axons comes from the many interactions of the water molecules with their membranes. In the direction parallel to the length of axons, there is no membranal interference and the diffusion is rapid.

A DANCE OF SPINES

A big question remains: what mechanisms can explain the swelling of neurons during their activation? For the time being, we can propose only hypothetical scenarios that experimental studies will have to confirm. The classic (mathematical) model of neuronal activation dates from the 1950s. For that model, Alan Hodgkin and Andrew Huxley received the Nobel Prize in

1963 with John Eccles. The cells of an organism at rest have a clear electrical charge, or potential, on their surface. Neurons do, too, but they are more excitable: when the potential is changed locally beyond a certain base value, a sort of "storm" is produced. Some channels open, in particular the sodium channel, and a large number of sodium ions penetrate into the neuron, which accentuates the depolarization of the membrane and provokes the opening of other sodium channels. This creates an explosive chain reaction that ends by reversing the potential of the membrane. At that moment, the channels close and the neuron recovers its base potential by allowing the potassium ions to escape, then by expulsing the excessive sodium ions with the pumps mentioned earlier.

The passage of ions through the membrane in one direction then in the other produces an electrical current in the neuron. The action potential is in fact propagated gradually along the membrane of the neuron (it goes in only one direction, the membrane remaining "struck down," and cannot be depolarized for a very short moment after the passing of the action potential, blocking any return backward). Each neuron "captures" signals coming from other neurons from multiple prolongations, dendrites, on which are found a large number of "spines," the ends of the axons coming from thousands of other neurons (figure 6.5). Each of these endings thus contributes more or less to depolarizing the membrane of the body of the neuron in function of the data transported by other neurons. Beyond a certain threshold of depolarization, the action potential is generated.

The potential is propagated from the body of the neuron along its axon ending in its extremity at the spine of another neuron, which it depolarizes in turn. This depolarization occurs by relegating neurotransmitters, such as acetylcholine or dopamine, to the extremities of the neurons of chemical molecules. Neurons are in fact in contact on their extremities, via a "synapse," the space that separates the axonal ending and the dendritic spines. The action potential liberates the neurotransmitter—stored up in the small vesicles on the extremities of the axon—in the synaptic gap. The neurotransmitter binds to a specific receptor of the dendritic spine and provokes the depolarization of its membrane, and so forth . . . The Spanish neuroscientist Santiago Ramon y Cajal discovered that neurons are organized in a contiguous and not continuous way through synapses. The artistic renderings he did are magnificent, and are still admired today. Cajal received the Nobel Prize in

1906 for his work with the Italian Camillo Golgi (who nevertheless did not believe Cajal's theory at all . . .).

Hodgkin's and Huxley's model explains neuronal functioning very well. But an increasing amount of data suggest that it doesn't explain everything. Thus, from the observed changes in potential, we can deduce the number of ions that have passed through a membrane. We know the number of water molecules associated with each of these ions from the electrical charges present, which is on the order of a few water molecules per ion, and this enables us to estimate the number of water molecules that must have passed through the membrane to conserve that balance. Recent work by the team of Pierre Magistretti and Pierre Marquet in Lausanne has shown that a much greater number of water molecules enter neurons when they are activated, on average 400 to 600 for each ion, which leads to a swelling of the neuron on the order of 6 to 7 percent (reference 6.22). Furthermore, these water movements are perfectly synchronous with the movements of ions (a very rapid entrance following by a slow expulsion), and those of the membrane observed directly by the Japanese biophysicist Ichiji Tasaki.[1] Tasaki's work, like that of Magistretti's team, shows in addition that the entrance of water into the neuron is accompanied by a change in density (and by an indication of optical refraction), which suggests a modulation of the structure of water. This temporary swelling (a few thousandths of a second) is, however, not homogenous over the entire neuron. Dendrites, especially at the level of their spines, the place where depolarization occurs, are particularly involved.

Considering the thousands of dendritic spines with which each neuron is endowed, and the very large number of neurons present in each column of the cortex, we must imagine the dynamics of the whole with the micro-swellings and shrinking of all these elements. Cerebral tissue is certainly not static, but is animated permanently by microscopic movements, which today are revealed in films created from tissular preparations observed under the microscope. Had Cajal guessed this while speaking of neurons as "mysterious butterflies of the soul whose beating wings could—perhaps—one day clarify the secret of mental life?" Francis Crick, who with James Watson (Nobel Prize in 1962) discovered the structure of DNA, at the end of his life had also

---

[1] Tasaki was a tireless researcher whom I had the honor of meeting at the NIH. He died in January 2009 at the age of 99. Up until his death he continued to work and to publish articles, walking the mile that separated his home from his lab every day.

begun to foresee the phenomenon in one of his last articles devoted to the "twitching of the dendritic spines" (reference 6.23). These are thus probably the incessant movements, increased or diminished depending on cerebral activation induced by various stimuli or conditions, which is revealed by diffusion MRI via the interaction of water molecules with membranes.

Mechanical Neurons

These mechanical effects seem to originate in a thin layer close to the membrane that contains many calcium ions and a thick matrix of filaments and microtubules very rich in negative charges (and thus likely to polarize water molecules): the cytoskeleton. Researchers have managed to upset this network of microtubules, even to make it soluble with chemical agents. The result: the action potential disappears. The cytoskeleton is thus indispensable to the functioning of the neuron. It is made up of proteic strands very rich in negative charges and maintained two-by-two thanks to the bridges of calcium ions that possess two positive charges, one for each strand. The whole is very compact due to a cooperation between the sites and the bridges. We can thus consider the cytoskeleton as a sort of compressed spring. When the neuron is depolarized, the flow of sodium ions begins to compete with the calcium ions. Since sodium ions have only a positive charge, they can't serve as bridges to maintain the elements of the cytoskeleton: it is sprung, like a released spring (the elements repel each other due to their negative charges), which makes the cell swell, thereby creating an appeal to water molecules that encounter the negative charges exposed by the unfolded strands of the cytoskeleton and the cellular membrane. A less dense network of water, with slow diffusion and diminished index of optical refraction, appears temporarily, increasing the pool of slow diffusion water. When the cytoskeleton can grow no longer, the process stops, at the peak of the action potential. Things then return to normal, the sodium ions are expelled from the cell, and the cytoskeleton recovers its initial compact form. The variation in charges exposed by the cytoskeleton and the variations in the structuring of the network of water molecules contribute to the change in potential of the membrane corresponding to the action potential.

The idea that the structure of the water in cells can vary and that the dynamics of water molecules can be associated with the potential of the membrane of neurons is not new. Already in 1961 Linus Pauling, mentioned earlier for his work on the magnetic structure of hemoglobin, had proposed a general theory to explain the anesthetic action of xenon and of certain molecules. Xenon, an inert gas, does not form hydrogen bonds, but is inserted in microdomains formed by water molecules, sorts of watery crystalline ministructures. We are also aware of the temporal asymmetry of the action potential and the change in water diffusion observed through MRI. The first phase, bound to the entrance of sodium ions and the unfurling of the cytoskeleton is very rapid, and doesn't consume energy, which is stored in the cytoskeleton and liberated during this phase. The restoring of the system expends a lot of energy and takes a bit more time. It is somewhat like a teenager's room: the disorder has a tendency to appear very quickly and without effort. By contrast, cleaning it up requires energy . . . and time! We can also begin to foresee why the increase in blood flow (upon which, let's recall, the BOLD functional MRI method rests), believed to bring energy to the activated tissue, need not be very rapid. It is not the neuronal activation that would consume the energy, but the restoring of the system after activation.

Another physical phenomenon, suggested by the laws of thermodynamics, must necessarily accompany this cellular swelling. It is well known that to make ice melt, it has to be heated, it needs energy. By contrast, when the water of a lake freezes, heat is produced. Orange growers in California know this well. When frost threatens oranges, which are very sensitive, growers sprinkle the ground with water. Upon freezing, the water emits heat, which protects the oranges. This production of heat, here too, is bound to the structuring of the network of water molecules. The more structured it is, the less it needs energy to maintain its cohesion. The structuring of a large number of water molecules around the membrane and the cytoskeleton that occurs during the first phase of the action potential must thus liberate heat. In other words, not only does the neuron swell, but it must also heat. Only a part of this heat (45 to 85 percent) is reabsorbed during the phase of restoration of the action potential, as thermal exchanges always have an irreversible side (like in a steam engine).

This heat production has been observed by several groups of researchers, and Tasaki has shown that it is completely synchronous with the action potential. The detected elevation in temperature is small, 23 microdegrees (reference 6.24), but considering that millions of neurons are activated at the same time, the heating can be sufficient to be harmful to neurons that are very sensitive to it (which would explain the convulsions that sometimes occur in febrile newborns). And water, as we have seen, has a very high caloric capacity. Might the increase in blood flow that accompanies cerebral activation thus occur quite simply to eliminate this heat and cool the activated neurons (the blood flow can maintain the temperature to a tenth of a degree)? This would explain that an increase in blood flow does not need to be very rapid or very localized, as we noted at the beginning of this chapter. And comparing a vascular response with the fire hydrant and the arrival of firefighters is perhaps not so far from reality . . . Ancient Egyptians had already noted that the brain heated up, and so they considered it to be a mere radiator and took it out during mummification, whereas other organs were preciously preserved. The hypothesis of the thermoregulatory role of blood flow is obviously rather controversial, and I sometimes like to tease my friend Seiji Ogawa by telling him he should have called his BOLD functional MRI method "COLD" instead . . .

But we can be even more provocative, challenging the ubiquity of the synaptic transmission in the functioning of neurons, which is now universally accepted. Physical laws are generally reversible. If the depolarization of neurons is accompanied by cellular swelling, the induction of a sizable "mechanical" change in the neuron should provoke a change in potential, just as in the piezoelectrical effect of a quartz crystal in our watches or in our lighters.[2] Are neurons truly sensitive to these mechanical effects? This is the question I asked one evening of Pierre Magistretti, my neurobiologist friend from Lausanne who was visiting Kyoto. The next day, right in the middle of his lecture, he suddenly stopped. He then described the experiments he was currently carrying out in Lausanne, and in particular how he was activating neurons by giving them a little shock via a micropipette. He repeated this and looked at me: thus, without realizing it, he had, like many other neurophysiologists, mechanically activated the neurons in his experiments! Neurons can thus be considered piezoelectric systems, and this coupling that I've called "electromechanical" could play an important role in the propagation of information in neuronal networks.

---

[2] Piezoelectricity is the property that certain bodies possess in order to be electrically polarized under the action of a mechanical constraint, and vice versa, to be deformed when an electrical field is applied to them, the two effects being indissociable.

Synaptic transmission is a slow phenomenon: the action potential must migrate along the axon at a relatively low speed. Then, the vesicles must open, and cast off their neurotransmitter that must then diffuse into the synaptic gap to reach the other side of the synaptic spine and depolarize the following neuron. Several thousandths of seconds are necessary. Electromechanical coupling is almost instantaneous. One part of the neuron swells (dentritic spines, for example), and cellular elements that are in contact "feel" it immediately, like the movements of a person in a packed subway car are felt by those standing next to him. We can thus imagine that within groups of neurons in direct contact (we speak of "gap junctions") information circulates very rapidly through this electromechanical process. By contrast, when the information must be transported at a distance, even a few millimeters, to reach another group of neurons, the synaptic transmission becomes indispensable. These hypotheses are, of course, still very speculative, even provocative, and still far from being accepted by the scientific community. Let's see how MRI systems might evolve in order to validate them, even partially.

SEVEN

THE CRYSTAL BRAIN

We are almost at the end of our voyage to the center of the magnetic brain. Up to now, it has been magnetized water that has served as our guide. But water molecules sometimes need a little help, even when they pass the baton to other molecules. Water is everywhere, but this can be a disadvantage when we attempt to localize it for imaging purposes, especially when it moves around through diffusion. And, of course, there are other interesting molecules in our brain.

Doctors in particular would like to know the nature of the lesions they see in images. The slowing of water diffusion in an ischemic region of the brain (in the case of a stroke) or near a cancerous legion is already very precious information. But can we detect that lesion if it is very small? Or, in the case of cancer, can we know its type and stage? Or can we see an atheroma plaque on the wall of a carotid artery before it detaches and causes a stroke?

It would be ideal to have a specific tracer for a given lesion, one that would go find a particular place or lesion and stick to it, telling us: "Here it is, I found what I was looking for." Somewhat like a trained dog in an airport that sits next to a suitcase indicating that it contains drugs. Such tracers are beginning to appear, enabling "molecular imaging." This involves complex and miniscule particles from the world of nanotechnology (figure 7.1). First, these particles have a molecular structure that specifically recognizes the components of normal (the wall of a blood vessel, for example, as these components are very numerous and abnormal in cancerous tissue), or abnormal tissue (inflamed cells, Alzheimer plaques); and second, a small "light" makes them visible in MRI. This can involve iron atoms that disturb the magnetization of the water molecules around them, or manganese and "rare earths," such as gadolinium, seen in chapter 3, which accelerates the speed at which water returns to its magnetization of equilibrium after having been disturbed by the radio waves of MRI (relaxation).

ENTER THE NANOPARTICLES

These tracers are already used in animals, for example, within the framework of functional MRI. The BOLD method is very powerful, as we have seen, but the changes in magnetization of water through our thoughts, either conscious or unconscious, remain tiny, detectable only with very sophisticated image analysis software. The BOLD principle rests on variations in the concentration of deoxygenated hemoglobin induced by variations in the blood flow. Hemoglobin is abundant, "free," but its magnetic moment (owing to its single iron atom) is not very high. To amplify the effect, some researchers have experimented with replacing it in fMRI with a much more effective nanoparticle, although that particle is necessarily much less abundant (it would have to be injected) and much more expensive (it must be manufactured and purchased). This nanoparticle has to contain a very large number of iron atoms that interact among themselves in synergy (in order to produce a "superparamagnetic" effect that greatly increases the magnetic moment [figure 7.2]) and must be "furtive" in order not to be caught too quickly in the reticuloendothelial system (which includes the liver, the spleen, and the lymph glands; they collect particles that circulate in the blood like flies on flypaper), so they will stay in circulation long enough, at least for the time it takes for the experiment. Among the ultra-small (nano) particles with iron oxide (USPIOs), ferumoxytol meets these criteria. Its chemical formula is $Fe_{5874}O_{8752}C_{11719}H_{18682}O_{9933}Na_{414}$, which means that it contains 5,874 iron atoms in the form of iron oxide hidden in an armor of "velvet" (the rest of the nanoparticle, or the "polyglucose sorbitol carboxymethyl ether").

Hemoglobin with its single iron atom isn't heavy enough . . . in every sense of the term (a ferumoxytol particle is 44 times heavier). Once injected in the blood, it remains in stable concentration for hours. In the brain, in activated regions, the associated increase in blood volume locally increases the quantity of nanoparticles in the small vessels, which modifies the local magnetic field and significantly *reduces* the signal coming from the water (around 3 times more) than the BOLD method *increases* the signal (because the increase in the blood flow causes the quantity of nonoxygenated hemoglobin to diminish) (figure 7.2; reference 7.1). These studies have been carried out on animals, but ferumoxytol has recently been authorized in the United States as a medicine to treat certain types of anemia and other associated iron deficiencies (we can understand why, given the enormous amount of iron present in the particle . . .). Some researchers already imagine its use for fMRI in humans.

Another interesting tracer in animals is manganese. This atom is similar to calcium in size and electrical charge, tastes like calcium (at least to neurons), but it is not calcium. It can therefore not activate neurons; but they love it, and capture it when they are activated through the calcium channels of their membranes. What is more, they siphon it into their axons and pass it to their neighbors via synapses! When a region is activated, the implicated neuronal system is charged in manganese ions, which accumulate and become visible in MRI because, like gadolinium, they accelerate the relaxation of neighboring water molecules. This method has two very great advantages over the preceding one (and the BOLD method). First, one sees directly the activation of neurons, and not the indirect effects of activation on the flow or volume of blood. Second, one does not need MRI for manganese to enter neurons. MRI intervenes only to reveal the presence of manganese in the activated circuits.

Using this process, Dr. Annemie Van der Linden in Antwerp has studied the physiological system that controls singing in birds. It is obvious that once the birds are placed into the magnet, anesthetized so they don't move (or simply fly away), we mustn't hope to hear them sing. On the other hand, they can sing to their heart's content in their cages, which causes the manganese atoms that had been administered to them earlier to pile up in the neuronal circuits involved. The singing is ephemeral, but the manganese remains. One need only make them sleep later and put them in an MRI scanner to reveal the circuits where the manganese has accumulated. This is how we were able for the first time to see how circuits evolve in function of genetic, hormonal, or seasonal influences (reference 7.2). Similar studies have dealt with the auditory and olfactory systems of rodents (reference 7.3), showing how the localization of subregions of the olfactory bulb differs depending on stimuli. Another application of manganese is for the visualization of intracerebral connections: it in fact "travels up" the length of the axons. Unfortunately, manganese is toxic and it can't be used in humans. Diffusion imaging, which requires no injection, thus remains the method of choice.

Within the framework of medical applications, molecular imaging has the potential to show us the regions where a given biological process is unfolding—neogenesis: the production of small blood vessels in cancerous tumors; apoptosis: the cellular death in neurodegenerative diseases; inflammation in multiple sclerosis or cerebral infarction, and so on. It can also indicate the expression of certain genes (which is of great interest in genetic therapy) or even the migration of cells in organs or tissue, that of stem cells, for example.

Thus complex molecular edifices have been conceived, rich in hydrogen atoms that are rapidly exchanged with those of water via hydrogen linking, and which exchange the magnetization that they have received in certain conditions. This is the chemical exchange–dependent saturation transfer (CEST) effect that can be rather specific for the presence of interesting biological molecules (reference 7.4). The "europium-DOTA-tetraamide" molecule, for example, which has "arms" in bisphenylboronate, has a CEST effect dependent on the presence of glucose (sugar), which potentially enables us to determine its concentration. Other molecules are sensitive to lactate (a reflection of the metabolic status of the cell), or even to the acidity of the environment, important parameters that describe the good (or bad) functioning of tissular cells. We even speak of "intelligent" tracers able to transform themselves when they encounter well-defined local conditions.

The tracer is thus invisible in the administered form, but it becomes visible when it is transformed. For example, we can be dealing with a complex molecule that contains a gadolinium atom, but that remains hidden, invisible to the water molecules that cannot get close to it. The molecule is in fact sensitive to an enzyme present in some cells, beta-galactosidase, which "digests" certain sugars. This enzyme indicates genic activity as seen in cells—in particular, in senescent cells in which it is abundant—or even because the gene corresponding to its production has been associated with another gene (injected through genetic manipulation) whose cellular expression we wish to verify (figure 7.3; reference 7.5). The enzyme splits a part of the molecule that resembles a sugar, allowing water molecules to approach and "see" the gadolinium. These molecules lose their magnetization (relaxation) and contaminate the other water molecules by exchanging themselves for the water.

A large number of water molecules will thus react, to the point of modifying the MRI signal, revealing the presence and the genic activity of the cells. We can also cause an abundant production of the transferrin receptor (a natural molecule transporting iron) on the surface of cells through genetic manipulation (by injecting through a viral vector a corresponding gene coupled with a gene of interest whose expression we wish to see). Nanoparticles very rich in iron (USPIO) that also contain an analog of natural transferrin, when injected, are intercepted by these receptors, internalized in the cell and concentrated in the vesicles. The magnetization of the local water molecules are modified and the cells expressing these genes are detected (reference 7.6). In addition, we can, again through genetic manipulation with a viral vector, synthesize in large quantities through cells a protein akin to ferritin, which couples with iron. These molecules will then pump the natural iron atoms from the organism and collect them in large quantities: in short, the cells work for us, producing the tracer themselves (figure 7.3; reference 7.7)! The cells (neurons, glial cells from the brains of mice) remain marked and visible in MRI for weeks, which enables us to follow their migration in the brain. We can also trace using stem cells and observe their fate after they have been injected into the brain.

MANY ARE CALLED, FEW ARE CHOSEN

All these examples highlight the potential of joining together molecular biology, genetics, and nanotechnologies. However, for the moment we have only "proofs of concepts" obtained through animals, or even sometimes only in tissue cultures. Molecular imaging through MRI in fact raises two types of problems. The first is that the tracer must arrive in sufficient quantities at its target. In general it is administered through injection and is transported in the blood. It cannot be toxic, obviously, which imposes very strong constraints on its nature and composition. After escaping the liver, which has a tendency to filter and eliminate that which is potentially toxic, and the network of the "reticuloendothelial system," the tracer must be able to enter the brain—that is, pass from the blood vessels into cerebral tissue. Now, the brain doesn't let much get through: the brain cells that surround the blood vessels are ferocious jailers that don't allow anything to filter through. This is what we call the "blood-brain barrier," a border indispensable to the functioning of our brain, one that is sensitive to tiny concentrations of molecules, especially if they resemble neurotransmitters.

The tracers must then be cunning enough to pass undetected, while remaining harmless. Sometimes a molecule "works" in an animal, and passes into its brain, but not in humans, who have a different type of blood-brain barrier. The molecule must then sight the target well, in the most specific (not mistaking its target) and the most sensitive way possible (so that as many of these molecules as possible reach a target in order to mark it). In addition, any molecules not fixed on a target must be rapidly eliminated by local physiological mechanisms (elimination in the blood) so as not to muddy the paths on the images. Thus, once the good idea—the principle of the tracer—has been established, thousands of molecular variants are synthesized and tested by producers of tracers in the hope that one of them "will work" in humans. Since humans come at the end of a series of tests, you can imagine the disappointment of teams in the event of failure. This is why the development of new tracers or drugs is so long and costly.

These issues of the biodistribution of tracers are not unique to MRI, but are shared with other imaging techniques that use tracers, including PET. By contrast, the place where PET is much better than MRI is on the level of sensitivity. In general, the action principle of the tracer in MRI is local perturbation of the magnetization of the water molecules, around the point of fixation of the tracer. Not only is the effect indirect, which implies that complex models must be used to go back to the source—the local concentration of the tracer, the only truly important parameter—but the signal is diluted because a considerable number of water molecules have seen nothing of the tracer, although they also contribute to the signal. Although it might not be exactly looking for a needle in a haystack, it's not very far from it. We must then inject a large amount of tracer to see something, which can raise issues of toxicity. With PET, on the contrary, the tracer alone produces a signal because it alone is radioactive. It is a bit like seeing fireflies at night. In addition, the tracer is seen directly and can easily be quantified (like the number of fireflies), even if sometimes complicated corrections are necessary.

So why should we bother with MRI? Because PET also poses a certain number of problems. The first is that we must "tag," or mark the tracer with a radioactive atom. This necessitates a cyclotron and a specialized chemical laboratory nearby. All molecules cannot be easily tagged. Above all, the radioactivity decreases rapidly, and the chemical syntheses must be very rapid while also having good results. Today, there are robots that achieve these syntheses very rapidly and effectively, but only insofar as a few molecules are concerned. With MRI, this problem of marking is also raised since one must integrate a visible atom for MRI (iron, gadolinium, and so on); but one has all the time in the world, which allows for the preparation of a large quantity of tracer in advance. This is impossible with PET, whose images, moreover, do not have the precision and resolution offered by MRI. This is why PET cameras are often hybrids, combined with an integrated x-ray CT scanner or even, recently, an integrated MRI imager. Since PET and MRI receiver signals are not really compatible, it has taken a great deal of imagination for builders to develop them. Moreover, one builder has proposed that the two systems not be combined, which would prevent the degradation of their performance: and so the PET camera and the MRI imager are separated by a few meters, but aligned. Thanks to a shuttle system, the patient undergoes his exam in one, then the bed on which he is lying slides on tracks, and he is ready to be examined in the other system. Rotating the patient between the two is even possible depending on whether it is necessary for the exam that he have his head or feet first.

It is clear that if we could be content with MRI and do without PET for molecular imaging, it would be much simpler and less expensive. What can we do to boost the less-sensitive MRI signal? Three paths are open to us: the strategy of the firefly, cooling, and rising to the summits of the magnetic field. The first path consists of finding atoms (or rather their nuclei) that are directly observable through MRI, without inducing an indirect effect on water, and that do not exist in a natural state in the organism, in order to avoid any "pollution" of the signal. This path is akin to that of PET, but without the use of radioactivity. Such a nucleus exists: fluoride. The concentration of fluoride in our teeth and bones is too weak and too immobile to be visible.

The nucleus of a fluoride atom has properties favorable to MRI. Synthetic compounds, such as fluorocarbons (which can be joined to oxygen and replace red blood cells) already exist, and some are anti-carcinogenic (fluorouracils). These compounds can be detected after having been injected, but do not have unique characteristics. We must then resort to new compounds, molecules rich in fluoride atoms and capable of recognizing particular targets. Promising attempts have been made in collaboration with our NeuroSpin laboratory by the French company Guerbet within the framework of the Franco-German Iseult project, which we will discuss later (see figure 7.4; reference 7.8). All the same, the minimal quantity of tracer detectable still remains a bit above that detectable through PET.

The second path opens when we look at the Boltzmann equations: clear magnetization is the result of the balance between magnetic moments parallel and antiparallel to the magnetic field. The number of magnetic moments carried by the molecules present in the tissue is considerable, but, as we have seen, thermal agitation to the temperature of the brain gives molecules a thermal energy on the same order of magnitude as the magnetic energy in the field of the magnet. The Boltzmann equations tell us that in these conditions the two orientations are almost equiprobable. Everything would be different if the temperature were lowered: the magnetic moments would be aligned and would all remain in the direction of the field, creating considerable magnetization. All the same, it is not a matter simply of cooling by a few degrees! It is in fact necessary to lower the temperature to close to absolute zero ($-273°C$), which makes this method difficult in studying the brain. Fortunately, there is a trick: hyperpolarization.

Hyperpolarization consists of "cooling" certain atoms (in a flask, not in the brain) to obtain a very intense magnetization of their electrons (reference 7.9). We force the magnetization of these electrons by various processes—for example, through optical pumping (a classic process using lasers) or by irradiating a solid rich in free electrons through microwaves (solid state dynamic nuclear polarization), all at very low temperatures. Then, this precious magnetization is transferred to the atomic nucleus that we want to study through MRI. The first attempts were made on xenon and helium. Even if the transfer of magnetization of electrons to nuclei shows weak results, the amount of magnetization and the very great number of nuclei present means that the magnetization obtained could be 100,000 times higher than that obtained in a standard MRI. This magnetization can be preserved for several hours by maintaining the nuclei in a small magnetic field, which enables them to be administered to an animal, even to a patient in an MRI scanner. For helium, a gas, one need only inhale it. We can thus obtain some very beautiful images of the air cavities, potentially interesting for pulmonary pathologies (emphysema, and so on). Time is of the essence, however, since the magnetization ultimately decreases.

It is important not to use standard MRI methods, however, which make use of radiofrequency impulses intended to "erase" the magnetization of the nuclei. This magnetization is then reformed (this is the principle of MRI), but at the level fixed by the value of the magnetic field of the scanner—all the advantages of hyperpolarization will have instantly been lost! We thus use radiofrequency impulses of very low intensity that involve only a small fraction of magnetization. As for xenon, it is potentially of use in imaging cerebral perfusion, but it would have a tendency to cause the patient to go to sleep (it is an anesthetic above a certain dosage).

Some research teams have succeeded in hyperpolarizing molecules that are more interesting biologically, such as pyruvic acid (which is an energetic substrate for cells). The hyperpolarized atom is carbon. Since natural carbon (^{12}C, with a nucleus of six protons and six neutrons) does not have a magnetic moment, we must use pyruvic acid synthesized through an isotope (non-radioactive) of carbon, carbon 13, which has seven neutrons. This makes things a bit more complicated and a bit more expensive, but that is not the principal obstacle. The problem is maintaining the hyperpolarization of the molecule long enough. For, once injected in the blood, the molecule loses its magnetization in contact with other molecules within a few seconds. It is thus necessary to act very quickly to acquire images. We can obtain images of the "vascular tree," the content of the blood vessels where the molecule is present, rather easily, but it is much more difficult to image the tissue: after injection, the molecule must have time to reach the targeted organ, then penetrate the tissue in a sufficient concentration, which takes time—in particular, in the brain, with its blood-brain barrier.

MRI of the Extreme

And so there remains the third path (again suggested by Bolzmann's equations). This time, the objective is to increase the magnetic energy of the atomic nuclei in relation to their thermal energy as much as possible. We must then increase the magnetic field of the MRI scanner. At 1.5 teslas, as we saw in chapter 1, the difference between "parallel" and "antiparallel" orientations is on the order of 50 nuclei of hydrogen per 1 million (at 37°C). At 3 teslas, we go to 100 nuclei per 1 million. We can imagine then that a true race has begun to achieve magnets with very "high fields" (7 teslas and more, or more than 140,000 times the Earth's field): to obtain the highest possible magnetic moment of the proton, nucleus of hydrogen, or other nuclei—this is the objective we hope to reach on a superior level as we continue on our conquest of the human brain.

On a different scale, this objective is somewhat akin to the guiding spirit behind the development and production of very large particle accelerators, like the Large Hadron Collider (LHC) at the Organisation Européenne pour la Recherche Nucléaire (CERN) in Geneva, to achieve increasingly high energy by making particles turn quicker and quicker, their trajectory in a ring of 27 kilometers being controlled by some 400 superconductor magnets. Out of this considerable energy, after frontal collisions with the particles, light bursts forth in the form of a particle that has been actively sought for a long time, the Higgs boson, a link that was lacking in our conception of physics and matter. Regarding MRI, in 2000 there were only two magnets in the world (or rather in the United States) operating at 4 teslas that were intended to be used on humans (one at the NIH, on which I had already done a few studies at the beginning of the 1990s with my friend Robert Turner), and one at 7 teslas in Minneapolis. A project for a magnet at 9.4 teslas was in the works in Chicago and two other magnets at 7 teslas were ordered, from Siemens (for the United States) and General Electric (for Japan).

There were many magnets operating at a higher field, 11.7 teslas, but only for use with animals. This is because it is not easy to produce such a field. In addition, the field must be extremely homogeneous (varying by less than one per million through the imaged object) and very stable (on the order of one part per billion during the acquisition of images). Over small dimensions, like the body of a mouse or a rat, this is imaginable, although the corresponding magnets can be extremely large, not much smaller than those required for the weaker magnets intended for humans. To achieve such fields, in fact, you must roll a conductor around an extremely thick cylinder.

This is the principle of the "solenoid," which we all remember from our school days: a copper wire is rolled in the shape of a cylinder and its ends are connected to the poles of a battery. The current circulating through the wire produces a magnetic field that can, for example, attract an iron nail. The same is true for MRI magnets, except that we are not attracting iron nails, but rather magnetizing the atomic nuclei of our brain. The current necessary to produce such intense fields is still enormous, on the order of several hundred amperes. At such intensity, a classic conductor—in copper—would be so overheated that it would very quickly melt. And so we call upon superconductivity, as we saw in chapter 1. The wire of MRI magnets is made not of copper, but of a niobium-titanium alloy, and the magnet is installed in a cryostat, a vat filled with liquid helium at a temperature of −269°C.

Is it possible to create a magnet operating at 11.7 teslas for use with humans? This was the challenge I proposed in 2001 to the CEA (Commissariat à l'énergie atomique et aux énergies alternatives/the French Alternative Energies and Atomic Energy Commission). Such a magnet should be considerably large, the laws of physics confronted with those of biology; the magnet should have the shape of a long cylinder to produce a homogenous spatial field, and although just the brain is to be examined, the whole body must be able to enter the magnet. A smaller magnet, just the size of a human head, would not be able to perform as well. The choice of 11.7 teslas (223,000 times the Earth's field) was motivated in part by the very promising results already observed in the brains of animals at this field, and also because it is the ultimate physical limit possible with a niobium-titanium alloy. Beyond 11.7 teslas other materials must be used, such as niobium-tin, which is much more expensive and difficult to manipulate.

At this higher field, images can have a resolution sufficient for us to observe the human brain on a scale that up to then has never been achieved, over a few thousand instead of a few million neurons. At the same time, for studies with animals (rodents), I suggested the acquisition of a 17 tesla magnet (340,000 times the Earth's field) in order to see the neurons almost individually. Thus the concept of a center for research in neuroimaging through intense-field MRI on the campus of Saclay, included on the campus of the CEA, was born. The idea was very well received, in particular by André Syrota, who was director of Life Sciences Department at the CEA at the time, and who was instrumental in raising the NeuroSpin project to the highest of levels. After a bit of brainstorming, "NeuroSpin" ("neuro" for the brain, and "spin," which represents the kinetic and magnetic moment of particles, a symbol of physics) was chosen as the name for this research center. It is the only platform in the world, at 11,000 square meters, that contains MRI scanners of unequaled power to explore the brain, of both animals and humans, with unprecedented precision, and that welcomes under the same roof experts in physics, biology, neurosciences, medicine, computer science, and signal processing, among others.

In addition to the know-how of the teams in MRI, neurobiology, and image-processing, we also wished to take advantage of the internationally recognized expertise of CEA physicists, engineers, and technicians in matters pertaining to magnets and "cryogenics" (the science of cold). These teams were indeed in the process of finalizing the delivery and installation of magnets from the LHC that they, themselves, had conceived. They were also involved in the development of confinement magnets for the International Thermonuclear Experimental Reactor (ITER), the international research center in Cadarache in the south of France that seeks to demonstrate the viability of producing energy through nuclear "fusion" (from hydrogen, as on the sun, as opposed to the nuclear "fission" at current centers that utilize a fuel with a base of uranium or radioactive plutonium). Since research on the human brain is indeed not less important than that into the production of energy by nuclear fusion or the quest of the Higgs boson, the task should appear just as noble to them, I thought.

A feasibility study was launched. Although Bruker Corporation accepted the challenge to produce the 17 tesla magnet for studies on animals, no one was ready to undertake the 11.7 tesla magnet intended for humans: too expensive, too long, performance below the projected expense. Only the CEA and the teams at DAPNIA (Department of Astrophysics, Particle Physics, Nuclear Physics, and Associated Instrumentation), today called IRFU (short for nothing less than Institute of Research into the Fundamental Laws of the Universe!) would accept the challenge of constructing a whole-body magnet 90 centimeters in diameter. The news spread very rapidly, and my colleagues across the Atlantic were amazed to see that in France we were launching into such a project, which they judged to be much too risky and ambitious, indeed impossible. I was called a madman.

After confirming the feasibility of the project with detailed studies, it was definitively validated by the administration of the CEA in 2004. All we needed was financing. A solution was found within the framework of a French-German partnership signed by President Chirac and Chancellor Schröder on April 30, 2005. It involved the association of manufacturers

(Siemens and Bruker in Germany, Guerbet and Alstom in France) and public institutions (CEA in France, and the University of Fribourg in Germany).
The objective was to conceive and manufacture not only a magnet but also
everything necessary to turn it into an MRI imager (with systems to produce gradients in the magnetic field, radiofrequency antennae, and so on).
Above all, this imager was to be subjected to an industrial process to establish
the feasibility of molecular imaging through MRI by using a magnetic field
as high as possible to increase its sensitivity and become competitive with
PET. The financing was provided half by the manufacturers, half from public
partners, in equal shares between France and Germany. It was the Agency
for Industrial Innovation (AII), later renamed OSEO, that was responsible
for financing on the French side.

A large part of this financing was intended for developments in molecular imaging (with targets such as Alzheimer's disease, inflammation in cerebral infarction, and brain tumors). The first stage, research and development
for the magnet and its production, took up around a fourth of the budget.
I proposed "Iseult" as the code name for the project, perhaps to reflect its
Franco-German character, as the tragedy has a French origin, but was put to
music by Wagner . . . That code name has remained on the French side, but
the Germans found it too romantic (Iseult dies at the end of the legend), and
proposed the acronym INUMAC (imaging of neurodisease using high-field
magnetic resonance and contrastophores), even though we pointed out to
them that it was a bit complicated and sounded like "inhuman" in French!

An Exceptional Instrument for an Exceptional Organ

The magnet is to be delivered to NeuroSpin in 2015. In the meantime, other
teams from around the world have not been sitting on their hands. In 2012,
around fifty MRI imagers operating at 7 teslas were installed throughout the
world (let's recall that there was only one in 2000, and many people didn't
see the point of them). There are four magnets for whole-body MRI scanners
(even if we're only imaging the head, we've seen the reasons for this), two in
Germany and two in the United States. A 10.5 tesla magnet is being installed
in the institution of my friend Kamil Ugurbil in Minneapolis, the home of
the intense field MRI. An 11.7 tesla magnet intended for humans, but measuring only 68 centimeters in diameter, was manufactured by the English

company Magnex (now gone) and installed at the NIH in 2011. Unfortunately, during the installation and testing stage this magnet has "quenched." Quenching is an event that occurs when the conductor at the heart of the magnet loses its properties of superconductivity. This can happen for various reasons: a lack of helium; local, unexpected heating beyond the nominal temperature (−269°C); mechanical vibrations; and so on. A small magnet is in fact very sensitive to interactions with other elements of the scanner, such as the gradient system that induces the currents and other considerable forces, whence the specification for the CEA magnet that it have a diameter of 90 centimeters.

Quenching is a dramatic event that must be avoided, but that unfortunately can occur. The wire, no longer being a superconductor, is transformed into a banal conductor that can certainly not hold the hundreds of amperes that circulate through it. A large quantity of heat is produced instantaneously, which immediately vaporizes the few hundreds or thousands of liters of liquid helium in the cryostat. The magnet is of course constructed not to explode, and a valve liberates a considerable volume of helium gas, still cold, giving the impression that it is snowing. Helium isn't toxic, but it is expensive, and the bill is high for replacing the thousands of evaporated liters. Above all, even if the magnet can survive quenching, there is all the same a high risk that it will be severely damaged.

Other countries, such as Japan, are currently considering manufacturing or acquiring magnets functioning at very high field for MRI (there exists a project for 14 teslas for humans in Korea . . .). Having seen my American colleagues recently, I asked them if they still thought I was crazy. They answered that I was still just as crazy as before, but that the problem was that I had become contagious . . .

The Iseult magnet of NeuroSpin will be very different from the others. There have been several innovative approaches (some are being patented) undertaken in collaboration with Guy Aubert, a world expert in magnetism, under the supervision of Pierre Védrine, head of the project for IRFU, with Franck Lethimonnier as the operational pilot for the entire project, in attempting to assemble this 120-ton mastodon (a classic 3 tesla MRI magnet weighs around 12 tons), whose development and production are regularly inspected, moreover, by a panel of international experts. Each step is rigorously validated, as needed, on functional "models" of the magnet, reproducing on a smaller scale and a much lesser magnetic field the final conditions of its operation.

The magnet's structure is not that of a solenoid, but of an assembly of 170 disks 2 meters in diameter with two spires, called "double wetted pancakes" (nothing to do with French crêpes). This design is original and has been mastered by the CEA teams. It is in a sense their signature, since it is found in the magnets the teams produced for the CERN. These disks are currently manufactured in Belfort in a hangar of the Alstom factories (next to those where the first French high-speed trains were built, as well as where the turbines for the hydraulic or nuclear electrical centers are today assembled), under the attentive eyes of the CEA physicists, rolling the "wire" with a precision on the order of a few dozen microns (figure 7.5). However, the disks will be subjected to considerable mechanical forces (around 8,000 tons) when the magnet will be "in field." The wire itself has been the object of great attention. Before arriving in Belfort, it is formatted in the United States by the manufacturer Luvata from a raw material (niobium-titanium alloy) from Japan following the specifications carefully laid out by the CEA teams.

This wire consists of an assembly of ten strands of niobium-titanium covered in copper, sitting within a copper duct of around 9 millimeters over 5. The total length for the magnet is 182 kilometers; the complete magnet having an external diameter and a length all told of a bit less than 5 meters. The nominal current circulating in the conductor will be on the order of 1,400 amperes, a level much greater than usual, and the energy present in the magnet will exceed 300 megajoules: we are getting close to the considerable energy stored in the magnets developed for nuclear fusion! Another unique aspect of our magnet is that it will be permanently fed by an external electrical source. The usual superconductor magnets for MRI are not, as the ends of the wire are connected to each other: no energy source is necessary, since there is no loss of energy in a superconductor wire with the current circulating *ad eternam* (as long as there is helium). The assemblage of double pancakes, each being connected to its neighbors, exposes it to small losses of current that must be compensated for by external feeding. This must be perfectly regulated to guarantee the extreme stability of the field necessary for MRI, which is by no means a trivial matter.

Two other important concepts have been developed. The first involves the "active shielding" of the magnet. In fact, the magnetic field produced by the magnet does not stop at its edges. It extends at a distance, although decreasing in strength. In order that it not interfere with all that can be found in its vicinity (data disks or magnetic cards that would be irretrievably erased), it is highly desirable to confine the magnetic field in the room where the magnet is found. The cylindrical arch 10 meters in diameter in which the Iseult magnet will be placed has been specially constructed. However, the field produced by the magnet will still travel a great distance if no precautions are taken.

A solution for isolating the magnet's field consists of surrounding the magnet with an iron shell. This is "passive" shielding. Thus, the 7 tesla magnet installed at NeuroSpin is in a 450-ton iron shell. For the Iseult magnet, it will take a shell of more than a thousand tons, which makes our architect (and our accountant) pale. "Active" shielding consists of producing a compensatory counterfield via a coil placed around the main magnet. This is the solution we've adopted for Iseult, with the shield acting as a secondary magnet also reinforcing the intensity and the homogeneity of the main field in order to reach the 11.7 teslas desired.

Another concept involves cooling the helium not at −269°C, the temperature threshold necessary for niobium-titanium to be a superconductor, but at −271°C—that is, just 1.8°C above absolute zero. At this temperature, liquid helium becomes a "superfluid," offering the astonishing spectacle of a fluid without viscosity climbing the walls of the receptacles in which it is poured, even jumping out as it produces a superb "fountain effect."

This property was observed in 1913 by the discoverers of superconductivity, but was unknown until 1937, the official date of the discovery of the phenomenon. The superfluidity of helium enables the instantaneous evacuation of heat, stabilizing the superconductivity of the niobium-titanium wire (fluctuations are permanently produced, with the risk of quenching if a critical threshold is crossed, all the easier if one is at the limit of the possible field for this alloy, which is the case here). Whereas for classic MRI magnets the 2,000 liters of helium are maintained through a system that recycles the vaporized helium, the volume and the constraints on Iseult's cooling system require the production of liquid helium on the spot (which is, moreover, a very good thing, as the costs of liquid helium have recently gone through the roof . . .). A true cryogenic factory to produce liquid helium has thus been installed in the basement of NeuroSpin, near the arch of the future magnet to minimize the loss of calories in the heat pipes connecting it to the magnet. The vaporized helium is recycled, collected in an enormous bladder 10 meters long, then pumped and compressed before being recooled to –271°C. All the elements of this cryogenic system are doubled in order to compensate for any defect that might threaten the magnet. Such is the Iseult magnet, a unique jewel of a scientific instrument.

THE BIRTH OF NEUROSPIN

When you have a jewel you need a case to put it in. The architecture of NeuroSpin was undertaken by Claude Vasconi, who conveyed his artistic vision onto the building (six arches representing the waves produced by the brain). The forms and colors of the building are lovely, with light dancing over the many glass surfaces depending on the hour of the day. However, there were many technical obstacles; it was essential that the various magnets not interfere with each other or with the laboratory materials. In addition to laboratories and offices for staff, the building was to contain space for preexam preparation for subjects, including patients, meaning we essentially needed a small hospital (with exam and training rooms, beds, and so on), and a place for preclinical research (electrophysiological training and recording rooms, biology and cell culturing), as well as other completely separate structures, both for research and for mechanical systems (air conditioning, for example) (figure 7.6).

The coordination of all this was conferred to Xavier Charlot, the head of the project for NeuroSpin, who toiled to meet the target date for the opening: January 2007. Today, NeuroSpin holds more than 150 people (25 percent of whom are international collaborators), working in four multidisciplinary laboratories on many unique projects with great potential: to push the current limits of MRI; establish the functional architecture of the brain on many levels; identify the mechanisms of cerebral development and plasticity, notably in children; establish the neural bases of the major cognitive functions (language, calculation, and so on) and even of consciousness; study the interactions between genes and the environment in the functional anatomy of the brain; research the analogies between animals and humans for diagnostic, and even "therapeutic" imaging (testing drugs by means of imaging), as imaging has become an extremely useful tool for drug companies, enabling them to see the effects of their molecules on the brain well before the appearance of clinical reactions.

For this research NeuroSpin has a range of imaging instruments (MRI of 3 to 7, and soon 11.7 teslas for humans, 7 to 17 teslas for animals, but also systems for recording the signals emitted naturally by the brain, electrical [EEG] or magnetic [MEG] with 306 captors placed around the head of the subject) and computers (networks, servers, calculators), which are essential in the processing of images (software specially developed for studying brain images, such as Brainvisa). This technology is of course available for all the NeuroSpin teams (which belong to the CEA and other research institutions, notably the Inserm [Institut national de la santé et de la recherche médicale/ National Institute of Health and Medical Research], INRIA [Institut national de recherche en informatique et en automatique/French Institute for Research in Computer Science and Automation], CNRS [Centre national de la recherche scientifique/French National Center for Scientific Research, or the Paris Hospital network], and so on), but is also open to the national and international community, public or industrial teams, depending upon the research these teams wish to carry out there. Such projects are often financed by prestigious national and international contracts and grants. The presence under the same roof of this singular technology and of renowned experts in very diverse fields, from fundamental physics to psychiatry, as well as computer science, mathematics, neurobiology, cognitive neuroscience, and radiology, not to mention electronics or mechanics, for which we have specific labs, makes NeuroSpin a truly unique place in the world. Clearly, our chances of one day understanding the inner mechanisms of cerebral functioning will be all the greater given all the different approaches that are being pursued together around imaging technology, as this book has attempted to describe.

An important part of the work in progress at NeuroSpin is in fact to prepare for the imminent arrival of the 11.7 tesla magnet. New technologies must be put into place to take maximum advantage of the capabilities of this unique magnet. The nominal frequency for MRI at this field will be 500 MHz (for the hydrogen nucleus), and this poses two important problems for the quality of the images and for safety. The first problem is linked to the fact that the wavelength associated with this frequency has an order of magnitude similar to the dimensions of the head. As a result there are phenomena of interference and stationary waves, which means that the field of radiofrequency is no longer homogeneous within the brain. It is a bit like our car radio: there can be a lot of static at a given place, and then the reception becomes normal again less than a meter away. The quality of the images suffers greatly, presenting white or black spots that have nothing to do with what is happening in the brain.

The solution consists of emitting MRI radio waves not with a single antenna, but with a series of antennae (called "coils") in a network located around the head. This technology has already been used with a much lower magnetic field to *receive* the MRI signal arriving in parallel from several regions in order to go faster, but now it is a matter of also *transmitting* the radio wave in parallel in a network of antennae. This means that we must send radiofrequency impulses into each element of the network precisely and in a controlled, differential way. These impulses must be calibrated precisely in function of the head of the subject. It then becomes possible to compensate for the spatial variations of the field of radiofrequency and to obtain images of excellent quality (reference 7.10).

Safety Above All

The second problem is that these radio waves heat up, all the more so when the frequency is raised (in function of the square of the field: between 11.7 and 3 teslas, the energy used increases by a factor of 15). Although the magnet is in no way inferior to those intended for nuclear fusion, it is of course out of the question to reach cerebral fusion! There are well-defined norms specifying the limits that cannot be exceeded. In general, the temperature of the brain must not be raised beyond one degree (the temperature rises more when you have the flu). MRI imagers are thus locked so that this limit is never reached, but this rarely happens at a lower field. At 11.7 teslas, that limit would often be reached (and exceeded) with the usual signal spatial encoding methods used in MRI. With the scanner being curbed, it would be almost impossible to obtain images, but this is where the imaging of parallel transmission enables us to overcome the obstacle. Instead of sending all the energy of the radiofrequency over the entire head once, only a small fraction is sent, and its temporal profile is optimized to homogenize the field of radiofrequency for each patient, with the energy from each antenna element (which, in addition, involves only a restricted spatial area) being individually calibrated through a precise modeling of the effects of heating. These techniques, developed and patented at NeuroSpin by Alexis Amadon and Martijn Cloos, will be essential for ultra-high-field MRI. Strategies are also being developed to monitor in real time the power transmitted through the antennae and the potential heating of the brain (reference 7.11), immediately cutting the power in the event of problems (for example, a defect in one of the antennae of the network).

This naturally leads to the question of the innocuousness of MRI scanners—in particular, at a very high magnetic field. Apart from the risk of overheating, there are two other potential risks. The first is the noise. As we have seen in chapter 1, producing images through MRI creates a loud noise, from the mechanical vibrations produced by the intense and rapid commutation of the currents in the gradient coils, in order to localize the signals. Providing auditory protection for the subjects is thus indispensable, but the noise generated increases with the intensity of the magnetic field. Thus, the noise risks being prohibitive for the Iseult magnet, unless it is limited to the study of deaf subjects (but, as we have seen in chapter 4, there are other auditory conduits than the ears . . .)! Here, too, clever strategies to limit the noise are being developed. First, new systems of gradients can be developed to generate less force, thus fewer vibrations. Second, encoding the MRI signal can be reviewed in order to solicit the commutation of gradients less. Some silent sequences have been developed by our colleagues in Minneapolis, with the acquisition of images being conveyed only by a discrete "pschiiittt"! However, it remains to be seen how this method can be used to generate all the necessary types of contrast.

And there is still the biggest question: how harmful is the magnetic field? The problem is that nothing has ever been reported, much less proven, and we can't really imagine what types of effects might appear. It is not like radioactivity, which harms DNA in particular, creating mutations that lead to cellular anomalies, and which can cause cancer or embryonic malformations. For MRI, we don't know where to look. Millions of MRI exams have been administered throughout the world every year for 30 years, without anything ever having been reported. However, there are indeed effects, some quite dramatic. The first is the "missile" effect. If a person approaches an MRI magnet without taking precautions (let's remember that with superconductivity the magnet is *always* in field, even when it is not being used, unlike what we are led to believe by a scene in a recent film where James Bond intervenes), his watch will be destroyed (transformed at best into a compass), his credit cards will be erased: nothing truly serious. But a pen with magnetic pieces, certain coins, keys . . . will be pulled out without warning and will go flying to the center of the magnet at incredible speed. If a person is in the way, the impact can be fatal.

Unfortunately, a few accidents of this type have occurred throughout the world, but the phenomenon remains very rare. The staff is of course extremely vigilant to verify that the people entering the room where the magnet is found are wearing no potentially dangerous objects. This involves all the staff—medical, technical, and of course maintenance: imagine the face of a cleaning woman seeing her pail fly off by itself! On the Web you can find photos of "collections" of various objects (a bottle of oxygen, wheelchair, stretcher . . .) that came a "bit too close" to the magnet (the force is not linear; the object leaves suddenly without warning) and were stuck to it, fortunately without anyone being in the trajectory. It takes extreme effort to remove those objects, and the magnet often has to quench for the field to disappear. Given this reality, it is easy to imagine that some people will *never* be able to have an MRI exam: those with pacemakers or neuronal implants or magnetic metal implants (manufacturers have made great strides to eliminate these, but they can still be present in older patients). In fact, any electronic prosthesis or magnetizable object in a body likely to be moved is prohibited. But a hip replacement or a metal plate on a bone do not a priori present any problems.

More subtle are the effects of the currents induced when one moves too quickly in the magnetic field, which decreases the farther one moves away from the magnet (in particular, near the entrance or the exit of the magnet). The laws of electromagnetism are universal and also apply to the ions in our cells: if we move in a field that varies in space, the ions in our tissue are subjected to forces and move in reaction. This tiny movement produces an electrical current, tiny as well, but which can depolarize some sensitive cells such as photoreceptors (the impression of seeing luminous flashes), receptors on the tongue (a metallic taste), or even the cells in the inner ear (feeling seasick). These effects are, however, completely harmless, and they cease immediately when one moves away from the field. They increase with a higher field (around one-third of people are sensitive at 7 teslas), but can be avoided with simple measures: slow, and no unnecessary, movements (like shaking one's head). For patients who remain in the central line of the magnet and who are then installed at the heart of the magnet where the field is the most homogeneous, there are in general no noticeable effects.

In the United States, the Food and Drug Administration has established recommendations for the use of MRI. Limited to 2 teslas at the beginning of MRI, fields up to 8 teslas are today considered harmless; this limit has evolved during the (short) history of MRI, depending on the number of machines available for use. Beyond 8 teslas, MRI can still be used, but after the advice of an ethics committee and a signature of informed consent by the subject, as is usual in all research protocols. In Japan, a manufacturer is marketing a 23 tesla magnet for NMR, and researchers do indeed have to get close to it to install samples (which are only very small test tubes containing chemical compounds from which they want to obtain NMR spectra, the diameter being much too small for imaging). However, a European legislation project could limit the magnetic field of MRI to much lower fields (in this case, it would be a law to be applied by all members of the Union, not a recommendation), from the principle of precaution, and this primarily for the "protection" of researchers and staff (patients are not considered by this law project). Let's remember that during the invention of the railroad, some thought that the body would not withstand a speed of more than 40 kilometers per hour. However, our European legislators traveling to Brussels around 300 kilometers per hour via the Thalys train obviously don't feel they are in much danger.

Among the four fundamental forces upon which physics rests, there is one that has considerable effects on our body: gravity. It can even have tragic effects if we fall from a tall building . . . However, no law prohibits the construction of buildings with multiple floors, and buildings are even becoming taller and taller. Limiting the use of MRI would obviously have dramatic consequences for health (patients would be deprived of an exam that has become the key to the vault for medicine, and the x-ray scanner, in spite of the risk of irradiation, would be used more) as well as on an economic level. It seems that that good sense will prevail, as this law project could be amended to exempt MRI at large. This does not, of course, prevent studying the potential effects of magnetic fields on cells, tissue, and organisms, as we are doing at NeuroSpin. Publications exist, but the related effects are most often not re-produced or verified, and are often contradictory, sometimes even anecdotal. Some Japanese researchers have been able to levitate frogs at the entrance of a 14 tesla magnet (the frogs came out unharmed, but were not transformed into princes and were not able to describe what they felt), others have repro-duced the miracle of Moses by causing the water contained in a receptacle to separate in two by the force of the magnetic field. But the extreme conditions necessary to obtain such effects are far from those that subjects who might agree to enter our 11.7 tesla magnet would encounter.

CONQUERING THE BRAIN

What are we hoping to see with this magnet? First, we want to see "better" what we already see with MRI imagers functioning at a lower field. The antic-ipated gain relative to the signal/noise should boost the sensitivity of MRI to detect the tracers developed within the framework of molecular imaging, the object of the Iseult project. We thus hope to observe lesions with unequaled precision, perhaps to detect Alzheimer's disease plaques in regions such as the hippocampus, where they appear very early.

The gain in signal will also enable us to study molecules other than those of water. These molecules, metabolites, which measure the energy level of the brain, and neurotransmitters, implicated in synaptic transmission, are very important in brain function, but are not easily detectable using standard MRI imagers because their concentration is very inferior to that of water. NMR spectroscopy enables us to detect their presence and to quantify them, but in relatively large volumes of cerebral tissue. At a very high field, we can expect to obtain images that show the extent of these molecules in the brain with a precision akin to that obtained with PET, but without having recourse to radioactivity. This advantage also enables us to acquire multiple images over a long period, for example during the carrying out of a cognitive task, in order to follow the evolution of those compounds.

But one of the major advantages of the research that can be carried out with this magnet is that we will be able to see what remains invisible today, and to discover things about which we as yet have no idea. We are experienc-ing a new phase of exploration; high-field MRI will enable us to set out to discover and conquer our brain just as spatial probes and large telescopes revealed the universe to us. And so at NeuroSpin, in collaboration with Luisa Ciobanu, we have already discovered, using our 17.2 tesla imager—unique in the world—that the effect of anesthetic agents on cerebral circulation is directly visible on images (even if they are of rats) and differ greatly depending on the anesthetic agents. The differences in contrast between blood vessels and cerebral tissue are directly linked to the levels of oxygenation of the blood in the brain: they reflect the impact of anesthetic agents on cerebral blood flow or metabolism, which must be minimal so as not to disturb the vital cerebral functions of the organism (figure 7.7; reference 7.12). This information was inaccessible up to now. Beyond the evaluation of new anesthetic agents and the understanding of little known effects of anesthesia on the brain, notably in humans, these results open a path to the study of disturbances in the cerebral circulation of blood associated with neurodegenerative diseases.

One of the great objectives for our 11.7 tesla magnet will be to discover whether a "neural code" exists, one underlying the general functioning of our brain, just like the genetic code that is responsible for the functioning of all our cells. Our knowledge is limited at present to the macroscopic scale (groups of several million neurons), which we observe in order to study our brain, whether we are dealing with regions activated by fMRI or connections revealed by diffusion MRI. We know that the Broca area is responsible for the production of language. However, we still don't know why and how this region processes language, except that functions other than language also seem to take place there. With fMRI we are thus in the process of establishing a catalog of cerebral regions or subregions with associated functions. Some regions seem rather specific, such as the region of "visual detection of words" that reacts in the presence of words when they are present in our visual field; but this region is very close to the one that recognizes faces, which is more developed among the illiterate. A French team has just shown that monkeys can also distinguish words made of real letters from words made of false letters, whereas a priori they don't know how to read. We have seen that the primary visual cortex was perhaps not so specific, since it also processes other information, auditory or tactile—in particular, among those who are blind at birth.

Evolution and the genetic code have given us a brain that has evolved, but very slowly, on the scale of a great many thousands of years. Cerebral plasticity must be coded in our genes, but the evolution of the functional anatomy of our brain during a life, or even in a few generations, cannot be of genetic origin. The ability to read is a recent (a few thousand years) acquisition of our brain, and the incredible dexterity with which adolescents use their cell phones or their video game controls has only been manifest for a few years. Is this a "recycling" of cerebral regions, available from the beginning to assume a given function, but that have deviated toward another function due to environmental pressures? Or are there stable and well-defined elementary functions, achieved in "blocks," low-level cerebral circuits, that are somehow joined together, connected, spatially and temporally, with other blocks to carry out more complex functions that appear continuously through our individual life or across generations?

In this case it would be the connectivity that plays the most important role in plasticity, as function, the fruit of evolution, is much less immutable over time on the scale of our life. We have around 20,000 genes. Granted, the combination of the expression of these genes in our cells presents many more possibilities, but our 100 billion neurons and the millions of billions of connections that they establish among themselves render the number of combinations almost infinite. The inventory of functional cerebral regions, as is possible today with fMRI, is a very important step, even if some have called it modern "phrenology" because it remains rather descriptive ("such a region is implicated in such and such a cognitive process"). Already, with the first magnets operating at high field, we have been able to see that a "region" (the resolution of images is at a few cubic millimeters at best) is sometimes constituted of a mosaic of regions that may have different functions.

At the other end of the spectrum, remarkable work accomplished over many years has demonstrated the individual "mechanics" of neurons on a microscopic and molecular level: we know a lot (but not all) about the biochemistry, the biophysics, and the molecular biology of the neuron. Electrophysiological research on animals enables the study of a few assemblages of neurons, but over very small regions. What we are lacking, then, is the intermediary realm, the one between the microscopic (1/100th of a millimeter) and the macroscopic (1 millimeter): the mesoscopic scale (1/10th of a millimeter). It is in the spatial and temporal arrangement formed by the assemblages of a few thousand neurons on a mesoscopic scale that we must seek the anatomical-functional specificity that makes up the brain, and perhaps the existence of an elementary "neural code." We must then be able to access it.

IN SEARCH OF A NEURAL CODE?

A number of disciplines, such as physics, are increasingly interested in this mesoscopic world, which is not the simple sum of the microscopic elements that comprise it. There are synergetic effects between the scales, causing new properties or functions to emerge, often in a nonlinear way, and that define complex systems. The brain most likely belongs here. The "addresses" of activated regions visible in fMRI are today often known by what we call the "Brodmann areas": the primary visual area is area 17, the Broca area of regions 44 and 45, and so on. These regions in fact correspond to an atlas published in 1909 by the German anatomist Korbinian Brodmann from his study of a few brains (there is even talk of a half-brain [hemisphere] of a grandmother, dissected in 1908 . . . Brodmann never provided the details of his work) (figure 7.8).

Brodmann had realized, by studying the different parts of the brain under the microscope, that the arrangement of the cells in the six layers that are generally found in the cortex (this number itself being variable) varies enormously from one site to another, transitions often being sharp. From this microscopic research he was able to identify fifty-two regions in the brain and establish the position of these regions on a typical brain. Today, the identification of regions activated in fMRI relies on this atlas, and images are redone to "jibe" with a typical brain for which the Brodmann areas were located in advance using spatial coordinates (x, y, z). We can imagine the rough nature of the approach (but we currently don't have anything much better): there is no reason for the "map" of the brain of a subject under study by fMRI to be the same as that of the half-brain of Brodmann's grandmother! These coordinates are used, rather, so that neuroscientists can agree on the position of the region they are talking about.

But these areas, defined by their cellular organization (we speak of "cytoarchitectony"), do indeed exist. They are also correlated with the microscopic organization of the nerve fibers that reach and leave them ("myelo-architectony") and the distribution of the neuromediators and neuronal receptors that are found in them, as is shown by studies of sections of the brain marked with radioactive tracers (figure 7.8; reference 7.13). On the functional level, the organization of the cortex in functional columns has been known for some 50 years, as was shown by the Americans David Hubel and Torsten Wiesel (Nobel Prize in 1981) for the organization of the visual cortex in a cat. Every half-millimeter, in humans, we find a group of thousands of neurons (columns of ocular dominance) that completely process visual information (orientation of elements, colors) coming alternately from each eye, right and left, which gives us the relief for a given point in space. This structure is reproduced identically on the surface of the visual cortex to cover all of the field of vision. The problem is that the organization of these regions, if it is globally the same in everyone, varies locally in some of us. The Brodmann areas, as they are used today in imaging, thus don't have true functional meaning; but they are a convenient means for spatial identification.

Very high-field MRI, with the spatial resolution that it will enable us to achieve, could offer us the path toward an individualized, true definition of the Brodmann areas. Above all, it will enable us to study in vivo and over the entirety of the brain (and not on sections) the spatial arrangement of cells, neurons, glial cells, and their connections. It will then be possible to identify the presence of functional groups, such as the columns of ocular dominance, but in other regions than the visual cortex. Just as the information on the genetic code is brought by the spatial arrangement of nucleotides, groups of atoms comprising the basic structure of DNA, we might imagine that a "neural code" is hidden in the spatial arrangement of neurons in the cortex, each spatial organization being responsible for a particular elementary function.

This microarchitecture and its associated functions would have a genetic programming—for example, at the origin of the Broca area (itself probably made up of many subregions)—but a global function such as language will emerge only when the connections with other functional elements, near or far, will have been established under the double effect of genetic programming (in particular, during the phase of cerebral development) and the environment. Such a neural code will thus enable an integration of the dual ability to respond to a genetic legacy and to be shaped by the environment, with the Broca area giving us the possibility to speak French, English, or Chinese.

REFERENCES

ONE

1.1—Schiller, F., *Paul Broca, explorateur du cerveau*, Odile Jacob, 1990.

1.2—Hounsfield, G. N., "Computed medical imaging," *Science*, 1980, 210, pp. 22–28.

1.3—Lauterbur, P., "Image formation by induced local interactions: Examples employing nuclear magnetic resonance," *Nature*, 1973, 242, pp. 190–191.

1.4—Kleinfeld, S., *A Machine Called Indomitable*, Times Book, 1985.

TWO

2.1—Dubois, J., et al., "Neurophysiologie clinique: développement cérébral du nourisson et imagerie par résonance magnétique," *Neurophysiologie clinique*, 2011, 42, pp. 1–9.

2.2—Jacquemot, C., et al., "Phonological grammar shapes the auditory cortex: A functional MRI study," *Journal of Neuroscience*, 2003, 23, pp. 9541–9546.

2.3—Molko, N., et al., "Brain anatomy in Turner syndrome: Evidence for impaired social and spatial-numerical networks," *Cerebral Cortex*, 2004, 14, pp. 840–850.

2.4—Martinot, J.-L., et al., "Brain morphometry and cognitive performance in detoxified alcohol-dependents with preserved psychosocial functioning," *Neuropsychopharmacology*, 2007, 32, pp. 429–438.

2.5—Maguire, E. A., et al., "Navigation-related structural change in the hippocampi of taxi drivers," *PNAS*, 2000, 97, pp. 4398–4403.

2.6—Gaser, C., and Schlaug, G., "Brain structures differ between musicians and non-musicians," *Journal of Neuroscience*, 2003, 23, pp. 9240–9245.

2.7—Keenan, J.-P., et al., "Absolute pitch and planum temporale," *NeuroImage*, 2001, 14, pp. 1402–1408.

2.8—Sowell, E. R., et al., "Mapping cortical change across the human life span," *Nature Neurosciences*, 2003, 6, pp. 309–315.

2.9—Draganski, B., et al., "Neuroplasticity: Changes in grey matter induced by training," *Nature*, 2004, 22, pp. 311–312.

THREE

3.1—Roy, C. S., and Sherrington, C. S., "On the regulation of the blood-supply of the brain," *Journal of Physiology*, 1980, 11, pp. 85–158.

3.2—Ter-Pogossian, M. M., "The origins of positron emission tomography," *Seminars in Nuclear Medicine*, 1992, 22, pp. 140–149.

3.3—Parsons, L. M., et al., "The brain basis of piano performance," *Neuropsychologia*, 2005, 43, pp. 199–215.

3.4—Belliveau, J. W., et al., "Functional mapping of the human visual cortex by magnetic resonance imaging," *Science*, 1991, 254, pp. 716–719.

3.5—Ogawa, S., et al., "Oxygenation-sensitive contrast in magnetic resonance image of rodent brain at high magnetic fields," *Magnetic Resonance in Medicine*, 1990, 14, pp. 68–78.

3.6—Turner, R., et al., "Echo planar time course MRI of cat brain oxygenation changes," *Magnetic Resonance in Medicine*, 2001, 22, pp. 159–166.

3.7—Kwong, K. K., et al, "Dynamic magnetic resonance imaging of human brain activity during primary sensory stimulation," *PNAS*, 1992, 89, pp. 5675–5679.

3.8—Ogawa, S., et al., "Intrinsic signal changes accompanying sensory stimulation: Function brain mapping with magnetic resonance imaging," *PNAS*, 1992, 89, pp. 5951–5955.

3.9—Bandettini, P. A., et al., "Time course EPI of human brain function during task activation," *Magnetic Resonance in Medicine*, 1992, 25, pp. 390–397.

FOUR

4.1—Le Bihan, D., et al., "Activation of human primary visual cortex during recall: A magnetic resonance imaging study," *PNAS*, 1993, 90, pp. 11802–11805.

4.2—Klein, I., et al., "Transient activity in the human calcarine cortex during visual-mental imagery: An event-related fMRI study," *Journal of Cognitive Neuroscience*, 2000, 12, pp. 15–23.

4.3—Thirion, B., et al., "Inverse retinotopy: Inferring the visual content of images from brain activation patterns," *NeuroImage*, 2006, 33, pp. 1104–1116.

4.4—Sadato, N., et al., "Neural networks for Braille reading by the blind," *Brain*, 1998, 121, pp. 1193–1194.

4.5—Merabet, L. B., et al., "Rapid and reversible recruitment of early visual cortex for touch," *PLoS One*, 2008, 3, p. e3046.

4.6—Bedny, M., et al., "Language processing in the occipital cortex of congenitally blind adults," *PNAS*, 2011, 108, pp. 4429–4434.

4.7—DeCharms, R. C., et al., "Control over brain activation and pain learned by using real-time functional MRI," *PNAS*, 2005, 102, pp. 18626–18631.

4.8—Meister, I. G., et al., "Playing piano in the mind—an fMRI study on music imagery and performance in pianists," *Cognitive Brain Research*, 2004, 19, pp. 219–228.

4.9—Hasegawa, T., et al., "Learned audio-visual cross-modal associations in observed piano playing activate the left planum temporale: An fMRI study," *Cognitive Brain Research*, 2004, 20, pp. 510–518.

4.10—Dehaene-Lamertz, G., Dehaene, S., and Hertz-Pannier, L., "Functional neuro-imaging of speech perception in infants," *Science*, 2002, 298, pp. 2013–2015.

4.11—Hertz-Pannier, L., et al., "Noninvasive assessment of language dominance in children and adolescents with functional MRI: A preliminary study," *Neurology*, 1997, 48, pp. 1003–1012.

4.12—Hertz-Pannier, L., et al., "Late plasticity for language in a child's non-dominant hemisphere: A pre- and post-surgery fMRI study," *Brain*, 2002, 125, pp. 361–372.

4.13—Krainik, A., et al., "Role of the supplementary motor area in motor deficit following medial frontal lobe surgery," *Neurology*, 2001, 57, pp. 871–878.

4.14—Kim, K.H.S., et al., "Distinct cortical areas associated with native and second languages," *Nature*, 1997, 338, pp. 171–174.

4.15—Somerville, L. H., et al., "Frontostriatal maturation predicts cognitive control failure to appetitive cues in adolescents," *Journal of Cognitive Neurosciences*, 2011, 23, pp. 2123–2134.

4.16—Pallier, C., et al., "Brain imaging of language plasticity in adopted adults: Can a second language replace the first?," *Cerebral Cortex*, 2003, 13, pp. 155–161.

4.17—Janata, P., et al., "The cortical topography of tonal structures underlying western music," *Science*, 2002, 298, pp. 2167–2170.

4.18—Blood, A. J., et al., "Intensely pleasurable responses to music correlate with activity in brain regions implicated in reward and emotion," *PNAS*, 2001, 98, pp. 11818–11823.

4.19—Dehaene, S., "How learning to read changes the cortical networks for vision and language," *Science*, 2010, 330, pp. 1359–1364.

4.20—Raizada, R.D.S., et al., "Socioeconomic status predicts hemispheric specialization of the left inferior frontal gyrus in young children," *NeuroImage*, 2008, 40, pp. 1392–1401.

4.21—Pinel, P., et al., "Genetic variants of FOXP2 and KIAA0319/TTRAP/THEM2 locus are associated with altered brain activation in distinct language-related regions," *Journal of Neuroscience*, 2012, 32, pp. 817–825.

4.22—Greene, J. D., et al, "The neural bases of cognitive conflict and control in moral judgment," *Neuron*, 2004, 44, pp. 389–400.

4.23—Knutson, B., et al., "Neural predictors of purchases," *Neuron*, 2007, 53, pp. 147–156.

4.24—Hesselmann, G., et al., "Spontaneous local variations in ongoing neural activity bias perceptual decisions," *PNAS*, 2008, 105, pp. 10894–10899.

4.25—Soon, C. S., et al., "Unconscious determinants of free decisions in the human brain," *Nature Neurosciences*, 2008, 11, pp. 543–545.

4.26—Dehaene, S., et al., "Imaging unconscious semantic priming," *Nature*, 1998, 395, pp. 597–600.

4.27—Owen, A. M., et al., "Detecting awareness in the vegetative state," *Science*, 2006, 313, pp. 1402–1403.

FIVE

5.1—Einstein, A., *Investigations on the Theory of Brownian Motion* (collection of papers translated from the German), ed. Furthe, R., and Cowper, A. D., Dover, 1956.

5.2—Stejskal, E. O., and Tanner, J. E., "Spin diffusion measurements: Spin echoes in the presence of a time-dependent field gradient," *Journal of Chemical Physics*, 1965, 42, pp. 288–292.

5.3—Le Bihan, D., and Breton, E., "Imagerie de diffusion in vivo par résonance magnétique nucléaire," *Comptes Rendus de l'Académie des Sciences*, Paris, 1985, 301, pp. 1109–1112.

5.4—Le Bihan, D., et al., "MR imaging of intravoxel incoherent motions: Application to diffusion and perfusion in neurologic disorders," *Radiology*, 1986, 161, pp. 401–407.

5.5—Le Bihan, D., et al., "Separation of diffusion and perfusion in intravoxel incoherent motion (IVIM) MR imaging," *Radiology*, 1988, 168, pp. 497–505.

5.6—Moseley, M. E., et al., "Early detection of regional cerebral ischemic injury in cats: Evaluation of diffusion and T2-weighted MRI and spectroscopy," *Magnetic Resonance in Medicine*, 1990, 14, pp. 330–346.

5.7—Turner, R., et al., "Echo planar imaging of intravoxel incoherent motions," *Radiology*, 1990, 177, pp. 407–414.

5.8—Warach, S., et al., "Fast magnetic resonance diffusion-weighted imaging of acute human stroke," *Neurology*, 1992, 42, pp. 1717–1723.

5.9—Takahara, T., et al., "Diffusion-weighted whole-body imaging with background body signal suppression (DWIBS): Technical improvement using free breathing, STIR and high-resolution 3D display," *Radiation Medicine*, 2004, 22, pp. 275–282.

5.10—Ross, B. D. et al., "Evaluation of cancer therapy using diffusion magnetic resonance imaging," *Molecular Cancer Therapeutics*, 2003, 2, pp. 581–587.

5.11—Iima, M. et al., "Apparent diffusion coefficient as an MR imaging biomarker of low-risk ductal carcinoma in situ," *Radiology*, 2011, 260, pp. 364–372.

5.12—Delannoy, J., et al., "Hyperthermia system combined with a magnetic resonance imaging unit," *Medical Physics*, 1990, 17, pp. 855–860.

5.13—Le Bihan, D., et al., "Temperature mapping with MR imaging of molecular diffusion: Application to hyperthermia," *Radiology*, 1989, 171, pp. 853–857.

5.14—Moseley, M. E., et al, "Diffusion-weighted MR imaging of anisotropic water diffusion in cat central nervous system," *Radiology*, 1990, 176, pp. 439–446.

5.15—Le Bihan, D., "Looking into the functional architecture of the brain with diffusion MRI," *Nature Reviews Neuroscience*, 2003, 4, pp. 469–480.

5.16—Douek, P., et al., "MR color mapping of myelin fiber orientation," *Journal of Computer Assisted Tomography*, 1991, 15, pp. 923–929.

5.17—Basser, P. J., et al., "MR diffusion tensor spectroscopy and imaging," *Biophysical Journal*, 1994, 66, pp. 259–267.

5.18—Basser, P. J., et al., "Estimation of the effective diffusion tensor from NMR spin echoes," *Journal of Magnetic Resonance*, 1994, 103B, pp. 247–254.

5.19—Pierpaoli, C., et al., "Diffusion tensor MR imaging of the human brain," *Radiology*, 1996, 201, pp. 637–648.

5.20—Poupon, C., et al., "Regularization of MR diffusion tensor maps for tracking brain white matter bundles," LNCS-1496, Springer-Verlag, MICCAI, MIT, 1998, pp. 489–498.

5.21—Mori, S., et al., "Three-dimensional tracking of axonal projections in the brain by magnetic resonance imaging," *Annals of Neurology*, 1999, 45, pp. 265–269.

5.22—Conturo, T. E., et al., "Tracking neuronal fiber pathways in the living human brain," *PNAS*, 1999, 96, pp. 10422–10427.

5.23—Tuch, D. S., et al., "High angular resolution diffusion imaging reveals intravoxel white matter fiber heterogeneity," *Magnetic Resonance in Medicine*, 2002, 48, pp. 577–582.

5.24—Wedeen, V. J., et al., "Diffusion spectrum magnetic resonance imaging (DSI) tractography of crossing fibers," *NeuroImage*, 2008, 41, pp. 1267–1277.

5.25—Hagmann, P., et al., "Mapping human whole-brain structural networks with diffusion MRI," *PLoS One*, 2007, 2, p. e597.

5.26—Klingberg, T., et al., "Microstructure of temporo-parietal white matter as a basis for reading ability: Evidence from diffusion tensor magnetic resonance imaging," *Neuron*, 2000, 25, pp. 493–500.

5.27—Dubois, J., et al., "Assessment of the early organization and maturation of infants' cerebral white matter fiber bundles: A feasibility study using quantitative diffusion tensor imaging and tractography," *NeuroImage*, 2006, 30, pp. 1121–1132.

5.28—Dubois, J., et al., "Microstructural correlates of infant functional development: Example of the visual pathways," *Journal of Neuroscience*, 2008, 28, pp. 1943–1948.

5.29—Bengtsson, S. L., et al., "Extensive piano practicing has regionally specific effects on white matter development," *Nature Neuroscience*, 2005, 8, pp. 1148–1150.

SIX

6.1—Friston, K. J., et al., "Analysis of functional MRI time-series," *Human Brain Mapping*, 1994, 1, pp. 153–171.

6.2—Turner, R., "How much cortex can a vein drain? Downstream dilution of activation-related cerebral blood oxygenation changes," *NeuroImage* 2002, 16, pp. 1062–1067.

6.3—Iwasa, K., et al. "Swelling of nerve fibers associated with action potentials," *Science* 1980, 210, pp. 338–339.

6.4—Latour, L. L., et al., "Spreading waves of decreased diffusion coefficient after cortical stimulation in the rat brain." *Magnetic Resonance in Medicine*, 1994, 32, pp. 189–198.

6.5—Zhong, J., et al., "Reversible, reproducible reduction of brain water apparent diffusion coefficient by cortical electroshocks," *Magnetic Resonance in Medicine*, 1997, 37, pp. 1–6.

6.6—Darquie, A., et al., "Transient decrease in water diffusion observed in human occipital cortex during visual stimulation," *PNAS*, 2001, 98, pp. 9391–9395.

6.7—Le Bihan, D., et al., "Direct and fast detection of neuronal activation in the human brain with diffusion MRI," *PNAS*, 2006, 103, pp. 8263–8268.

6.8—Miller, K. L., et al., "Evidence for a vascular contribution to diffusion FMRI at high b value," *PNAS*, 2007, 104, pp. 20967–20972.

6.9—Kohno, S., et al., "Water-diffusion slowdown in the human visual cortex on visual stimulation precedes vascular responses," *Journal of Cerebral Blood Flow and Metabolism*, 2009, 29, pp. 1197–1207.

6.10—Aso, T., et al., "An intrinsic diffusion response function for analyzing diffusion functional MRI time series," *NeuroImage*, 2009, 47, pp. 1487–1495.

6.11—Flint, J., et al., "Diffusion weighted magnetic resonance imaging of neuronal activity in the hippocampal slice model," *NeuroImage*, 2009, 46, pp. 411–418.

6.12—Yacoub, E., et al., "Decreases in ADC observed in tissue areas during activation in the cat visual cortex at 9.4T," *Magnetic Resonance Imaging*, 2008, 26, pp. 889–896.

6.13—Le Bihan, D., and Johansen-Berg, H., "Diffusion MRI at 25: Exploring brain tissue structure and function," *NeuroImage*, 2012, 61, pp. 324–341.

6.14—Dixon, W. T., "Separation of diffusion and perfusion in intravoxel incoherent motion MR imaging: A modest proposal with tremendous potential," *Radiology*, 1988, 168, pp. 566–567.

6.15—Le Bihan, D., "Intravoxel incoherent motion perfusion MR imaging: A wake-up call," *Radiology*, 2008, 249, pp. 548–552.

6.16—Niendorf, T., et al., "Biexponential diffusion attenuation in various states of brain tissue: Implications for diffusion-weighted imaging." *Magnetic Resonance in Medicine*, 1996, 36, pp. 847–857.

6.17—Vuilleumier, R., and Borgis, D., "Molecular dynamics of an excess proton in water using a non-additive valence bond force yield," *Journal of Molecular Structure*, 2006, 436, pp. 555–565.

6.18—Marx, D., et al., "The nature of the hydrated excess proton in water," *Nature*, 1999, 397, pp. 601–604.

6.19—Le Bihan, D., "The 'wet mind': water and functional neuroimaging," *Physics in Medicine and Biology*, 2007, 52, pp. R57–R90.

6.20—Ling, G. N., "The physical state of water in living cells and its physiological significance," *International Journal of Neuroscience*, 1970, 1, pp. 129–152.

6.21—Pollack, G. H., "The role of aqueous interfaces in the cell," *Advances in Colloid and Interface Science*, 2003, 103, pp. 173–196.

6.22—Jourdain, P., et al., "Determination of transmembrane water fluxes in neurons elicited by glutamate ionotropic receptors and by the cotransporters KCC2 and NKCC1: A digital holographic microscopy study," *Journal of Neuroscience*, 2011, 31, pp. 11846–11854.

6.23—Crick, F., "Do dentritic spines twitch?" *Trends in Neurosciences* 1982, pp. 44–46.

6.24—Tasaki, I., and Iwasa, K., "Temperature changes associated with nerve excitation: Detection by using polyvinylidene fluoride film," *Biochemical and Biophysical Research Communications*, 1981, 101, pp. 172–176.

SEVEN

7.1—Vanduffel, W., et al., "Visual motion processing investigated using contrast agent-enhanced fMRI in awake behaving monkeys," *Neuron*, 2001, 32, pp. 565–577.

7.2—Van der Linden, A., et al., "Applications of manganese-enhanced magnetic resonance imaging (MEMRI) to image brain plasticity in song birds," *NMR in Biomedicine*, 2004, 17, pp. 602–612.

7.3—Yu, X., et al., "In vivo auditory brain mapping in mice with Mn-enhanced MRI," *Nature Reviews Neuroscience*, 2005, 8, pp. 961–968.

7.4—Zhang, S., et al., "PARACEST agents: Modulating MRI contrast via water proton exchange," *Accounts of Chemical Research*, 2003, 36, pp. 783–790.

7.5—Louie, A. Y., et al., "In vivo visualization of gene expression using magnetic resonance imaging," *Nature Biotechnology*, 2000, 18, pp. 321–325.

7.6—Weissleder, R., et al., "In vivo magnetic resonance imaging of transgene expression," *Nature Medicine*, 2000, 6, pp. 351–355.

7.7—Genove, G., et al., "A new transgene reporter for in vivo magnetic resonance imaging," *Nature Medicine*, 2005, 11, pp. 450–454.

7.8—Giraudeau, C., et al., "High sensitivity 19F MRI of a perfluorooctyl bromide emulsion: Application to a dynamic biodistribution study and oxygen tension mapping in the mouse liver and spleen," *NMR in Biomedicine*, 2012, 25, pp. 654–660.

7.9—Viale, A., et al., "Hyperpolarized agents for advanced MRI investigations," *Quarterly Journal of Nuclear Medicine and Molecular Imaging*, 2009, 53, pp. 604–617.

7.10—Cloos, M. A., et al., "kT-points: Short three-dimensional tailored RF pulses for flip-angle homogenization over an extended volume," *Magnetic Resonance in Medicine*, 2012, 67, pp. 72–80.

7.11—Massire, A., et al., "Thermal simulations in the human head for high field MRI using parallel transmission," *Journal of Magnetic Resonance Imaging*, 2012, 35, pp. 1312–1321.

7.12—Ciobanu, L., et al., "Effects of anesthetic agents on brain blood oxygenation level revealed with ultra-high field MRI," *PLoS One*, 2012, 7, p. e32645.

7.13—Zilles, K., and Amunts, K., "Centenary of Brodmann's map—conception and fate," *Nature Reviews Neuroscience*, 2010, 11, pp. 139–145.

FIGURE CREDITS

Figures 1.1, 1.6, 2.3: Images reconstructed with BrainVisa software, courtesy of Y. Cointepas, D. Rivière, J. F. Mangin, © NeuroSpin/CEA.

Figure 1.2: Courtesy of E. A. Cabanis et al., Hôpital des Quinze-Vingts, Paris.

Figures 1.3, 3.3: Courtesy of S. Lehéricy et al., Hôpital de la Salpêtrière.

Figures 1.4, 1.5A, 1.5B, 2.1, 2.5, 3.1, 3.4, 3.6, 3.7 A and C, 4.2, 4.3, 5.1, 5.2, 5.3, 5.6 A and C, 6.1, 6.2, 6.4, 6.5, 7.2A: © D. Le Bihan.

Figure 2.2: © NeuroSpin/CEA.

Figure 2.4: Courtesy of J. Dubois, G. Dehaene, et al., NeuroSpin, and P. Huppi, Hôpitaux de Genève.

Figure 2.6: Courtesy of C. Poupon, et al., © NeuroSpin/CEA.

Figure 2.7: Reproduced with permission from Maguire, E. A., et al., "Navigation-related structural change in the hippocampi of taxi drivers," *PNAS*, 2000, 97, pp. 4398–4403. Copyright © 2000 National Academy of Sciences, U.S.A.

Figure 2.8: Adapted from Molko, N., et al., "Brain anatomy in Turner syndrome: Evidence for impaired social and spatial-numerical networks," *Cerebral Cortex*, 2004, 14, pp. 840–850.

Figure 2.9A: Reprinted with permission from Gaser, C., and Schlaug, G., "Brain structures differ between musicians and nonmusicians," *Journal of Neuroscience*, 2003, 23, pp. 9240–9245.

Figure 2.9B: Adapted with permission from Draganski, B., et al., "Neuroplasticity: Changes in grey matter induced by training," *Nature*, 2004, 22, pp. 311–312.

Figure 3.2: Reprinted with permission from Posner, M. I., and Raichle, M. E., *Images of Mind*, Scientific American Books, 1994.

Figure 3.5: © D. Le Bihan. Images (rat brain) reprinted with permission of Ogawa, S., et al., "Oxygenation-sensitive contrast in magnetic resonance image of rodent brain at high magnetic fields," *Magnetic Resonance in Medicine*, 1990, 14, pp. 68–78.

Figure 3.7B: © Natural History Museum, London.

Figures 3.8, 4.7: Courtesy of L. Hertz-Pannier and C. Chiron, CEA.

Figure 4.1: Reprinted with permission from Klein, I., et al., "Transient activity in the human calcarine cortex during visual-mental imagery: An event-related fMRI study," *Journal of Cognitive Neuroscience*, 2000, 12, pp. 15–23.

Figure 4.4: Adapted with permission from DeCharms, R. C., et al., "Control over brain activation and pain learned by using real-time functional MRI," *PNAS*, 2005, 102, pp. 18626–18631. Copyright © 2005 National Academy of Sciences, U.S.A.

Figure 4.5: Reprinted with permission from Hasegawa, T., et al., "Learned audio-visual cross-modal associations in observed piano playing activate the left planum temporale: An fMRI study," *Cognitive Brain Research*, 2004, 20, pp. 510–518.

Figure 4.6: Reprinted with permission from Dehaene-Lamertz, G., Dehaene, S., and Hertz-Pannier, L., "Functional neuro-imaging of speech perception in infants," *Science*, 2002, 298 (5600), pp. 2013–2015.

Figure 4.8: Reprinted with permission from Krainik, A., et al., "Role of the supplementary motor area in motor deficit following medial frontal lobe surgery," *Neurology*, 2001, 57, pp. 871–878.

Figure 4.9B: Reprinted with permission from Kim, K.H.S., et al., "Distinct cortical areas associated with native and second languages," *Nature*, 1997, 338, pp. 171–174.

Figure 4.10: Adapted with permission from Hesselmann, G., et al., "Spontaneous local variations in ongoing neural activity bias perceptual decisions," *PNAS*, 2008, 105, pp. 10894–10899. Copyright © 2008 National Academy of Sciences, U.S.A.

Figure 4.11: Adapted with permission from Soon, C. S., et al., "Unconscious determinants of free decisions in the human brain," *Nature Neurosciences*, 2008, 11, pp. 543–545.

Figure 4.12: Adapted with permission from Dehaene, S., et al., "Imaging unconscious semantic priming," *Nature*, 1998, 395, pp. 597–600.

Figure 4.13: Reprinted with permission from Owen, A. M., et al., "Detecting awareness in the vegetative state," *Science*, 2006, 313, pp. 1402–1403.

Figure 5.4: Courtesy of Dr. Koyama, University Hospital of Kyoto.

Figure 5.5: Courtesy of Dr. B. Ross, University of Michigan, Ann Arbor.

Figure 5.6B: Reprinted with permission from Douek, P., et al., "MR color mapping of myelin fiber orientation," *Journal of Computer Assisted Tomography*, 1991, 15, pp. 923–929.

Figure 5.7A: Adapted with permission from Mori, S., et al., "Three-dimensional tracking of axonal projections in the brain by magnetic resonance imaging," *Annals of Neurology*, 1999, 45, pp. 265–269.

Figure 5.7B: Adapted with permission from Conturo, T. E., et al., "Tracking neuronal fiber pathways in the living human brain," *PNAS*, 1999, 96, pp. 10422–10427. Copyright © 1999 National Academy of Sciences, U.S.A.

Figure 5.8: Courtesy of C. Poupon, B. Schmitt, A. Lebois, et al., © NeuroSpin/CEA.

Figure 5.9: Courtesy of J. Dubois, et al., © NeuroSpin/CEA.

Figure 5.10: Adapted with permission from Bengtsson, S. L., et al., "Extensive piano practicing has regionally specific effects on white matter development," *Nature Neuroscience*, 2005, 8, pp. 1148–1150.

Figure 6.3: Courtesy of Y. Fujiyoshi, Kyoto University.

Figure 7.1: Courtesy S. Mériaux, © NeuroSpin/CEA.

Figure 7.2B Reprinted with permission from Vanduffel, W., et al., "Visual motion processing investigated using contrast agent-enhanced fMRI in awake behaving monkeys," *Neuron*, 2001, 32(4), pp. 565–577.

Figure 7.3A: Reprinted with permission from Louie, A. Y., et al., "In vivo visualization of gene expression using magnetic resonance imaging," *Nature Biotechnology*, 2000, 18(3), pp. 321–325.

Figure 7.3B: Reprinted with permission from Genove, G., et al., "A new transgene reporter for in vivo magnetic resonance imaging," *Nature Medicine*, 2005, 11, pp. 450–454.

Figure 7.4: Courtesy of C. Giraudeau, Neurospin/CEA and Guerbet.

Figure 7.5: © Irfu/CEA.

Figure 7.6: © NeuroSpin/CEA.

Figure 7.7: Courtesy L. Ciobanu, et al., © NeuroSpin/CEA.

Figure 7.8: Reprinted with permission from Zilles, K., and Amunts, K., "Centenary of Brodmann's map—conception and fate," *Nature Reviews Neuroscience*, 2010, 11, pp. 139–145.

INDEX